Math Mammoth Kindergarten Worktext A

By Maria Miller

ISBN 978-1-954379-49-7

2026 Edition

Contents

Chapter 3: Addition & Measurement

Chapter 4: Subtraction

Foreword

Math Mammoth Kindergarten Workbook A and Kindergarten Workbook B comprise a complete math curriculum for the kindergarten mathematics studies. The curriculum meets the Common Core standards.

The main areas of study in these two workbooks are:

1. Basic counting skills within 0-100.
2. Addition and subtraction within 0-10, including word problems.
3. Identify, describe, analyze, and compose basic shapes.

Additional topics covered are sorting objects, comparing quantities and numbers, some foundations for measurement, position words, ordinal numbers, and counting coins.

This book, the kindergarten workbook A, covers counting and writing numbers (chapter 1), groups, comparing, and sorting (chapter 2), addition and measurement (chapter 3), and subtraction (chapter 4). The workbook B covers shapes, measurement, two-digit numbers, money, position words, ordinal numbers, and more addition and subtraction.

I heartily recommend that you read the full user guide in the following pages.

I wish you success in teaching math!

Maria Miller, the author

User Guide

Note: You can also find the information that follows online, at https://www.mathmammoth.com/userguides/ .

The Common Core Standards documentation is available at:
https://www.mathmammoth.com/preview/standards/MM-CCS-K-2026.pdf

Basic principles in using the Math Mammoth Kindergarten curriculum

Math Mammoth kindergarten curriculum is overall mastery-based, but it is slightly more spiral than the other grade levels of Math Mammoth, because most chapters deal with *two* major topics instead of one.

The kindergarten level differs from the other levels of Math Mammoth also in the fact that it is not expected that the child read the instructions or work independently very much. This level is meant to be **teacher-led**.

There are two student books (A and B). The chapters in them are intended to be studied mostly in order. However, you can do the lessons on shapes, measurement, and position words earlier or later, as you wish.

The **two student worktexts** (parts A and B) contain all the lesson material and exercises. Many of the lessons include activities that teach concepts, and those are explained right in the student books. Don't skip the activities, as they are an integral part of learning the concepts and skills for this level.

Some lessons include an explanation of the concept in a blue box. You, as the parent/teacher, need to explain the concept to the child(ren), using the blue teaching box as a guide. And then, almost all of the lessons contain some exercises for the child to do in the book.

Each chapter introduction explains the basic goals for the chapter. Most of them also list **games** (both online and offline games) that you can use to reinforce and practice the main skills in the chapter. Then, each chapter introduction also includes a link to a list of free online games and activities from other websites. All of these games can be used to supplement the math lessons, for learning math facts, or just for some fun.

We also include a versatile **worksheet maker** that allows you to make additional worksheets for the main topics in the curriculum. You will need Internet access to be able to use it. Access it here:

https://www.mathmammoth.com/private/Make_extra_worksheets_K.htm

Manipulatives

Below you will find a list of the recommended and optional manipulatives for the activities and games.

Recommended manipulatives

- a 100-bead abacus
- dice
- dominoes
- number cards, UNO cards, or regular playing cards
- a set of geometric solids (part B, chapter 7)
- a set of pattern blocks (part B, chapter 5)
- toothpicks and modeling clay (part B, chapter 7; for building geometric solids)
- a tangram set (part B, chapter 7)

Optional manipulatives

- counting bears (optional but can be fun because the sets often come with sorting cups and activity cards)
- math linking cubes (Unifix cubes)
- Cuisenaire rods (optional)

Many of the activities use various commonly found items that you probably have on hand in your home. For your convenience, we include a list of the recommended items and links to Amazon below. Note: You only need *one* item from each section. We're providing several options, plus naturally there exist many more choices for these products.

Recommended manipulatives

A 100-bead abacus or school abacus

There are lots of choices; search Amazon's collection here:
https://www.amazon.com/s?k=school+abacus&tag=mathmammoth-20

Learning Resources 10-Row Counting Frame or Abacus for Kids
This abacus has every five beads in alternating color (red, white), which is nice.
https://www.amazon.com/dp/B01LX6K2CV/?tag=mathmammoth-20

Wooden Abacus for Kids with Counting Sticks and Number Cards
https://www.amazon.com/dp/B08SBWKJZJ/?tag=mathmammoth-20

Geometric Pattern Blocks and Tangram-like games

Coogam Wooden Pattern Blocks Set 130PCS
https://www.amazon.com/dp/B07MYYK64R/?tag=mathmammoth-20

Beestech Wooden Tangram game with a puzzle Book
https://www.amazon.com/dp/B07Y37NNSR/?tag=mathmammoth-20

Vanmor Wooden Tangram set with shape cards
https://www.amazon.com/dp/B07XJ23KPG/?tag=mathmammoth-20

Wooden pattern blocks shape set with pattern cards
https://www.amazon.com/dp/B0CMTB3DR2/?tag=mathmammoth-20

GEMEM wooden pattern blocks set with design cards
https://www.amazon.com/dp/B07QPS5CFL/?tag=mathmammoth-20

Geometric solids (3-D shapes)

DEEKIN 36 Pcs Geometric Shapes $21
https://www.amazon.com/dp/B0BN8TXVK6/?tag=mathmammoth-20

DEEKIN 70 Pcs Wood Geometric Solids
https://www.amazon.com/gp/product/B0BC19RL8V/?tag=mathmammoth-20

Learning Resources Wood Geometric Solids Set of 19
https://www.amazon.com/dp/B000T1QQZ8/?tag=mathmammoth-20

Search Amazon's collection for geometric solids:
https://www.amazon.com/s?k=geometric+solids&tag=mathmammoth-20

Optional manipulatives

Cuisenaire Rods
https://www.amazon.com/dp/B016DU8UR0/?tag=mathmammoth-20

Counting Bears

Bmag counting bears
https://www.amazon.com/dp/B07ZNYDLXY/?tag=mathmammoth-20

Neorod counting bears
https://www.amazon.com/dp/B085ZSMK8P/?tag=mathmammoth-20

Linking Cubes

Learning Resources MathLink Cubes (100 pcs)
https://www.amazon.com/dp/B000URL296/?tag=mathmammoth-20

200 Pcs Math Linking Cubes
https://www.amazon.com/dp/B0GGCK8L8D/?tag=mathmammoth-20

Search Amazon's collection for math linking cubes:
https://www.amazon.com/s?k=math+linking+cubes&tag=mathmammoth-20

Pacing the curriculum

Each chapter introduction contains a suggested pacing guide for that chapter. You will see a summary on the right.

Most lessons are one page long, intended for one day. Some lessons are two pages, either intended for one day or two. There are also some optional lessons (e.g. about coins), not included in the charts on the right.

It can also be helpful to calculate a general guideline as to how many pages per week the student should cover in order to go through the curriculum in one school year.

Worktext A	
Chapter 1	19 days
Chapter 2	14 days
Chapter 3	25 days
Chapter 4	21 days
TOTAL	79 days

Worktext B	
Chapter 5	14 days
Chapter 6	17 days
Chapter 7	20 days
Chapter 8	13 days
TOTAL	64 days

The chart below lists how many pages there are for the student to finish in this particular grade level, and gives you a guideline for how many pages per day to finish, assuming a 170-day (34-week) school year. The page count in the table below *includes* the optional lessons (in particular, the lessons about coins).

Example:

Grade level	Lesson pages	School days	Pages to study per day	Pages to study per week
K-A	89	91	1	5
K-B	79	79	1	5
Kindergarten total	168	170		

The table below is for you to fill in. To get the count of "pages to study per day", **divide the lesson pages by the school days**. Lastly, multiply this number by 5 to get the approximate page count to cover in a week.

Grade level	Lesson pages	School days	Pages to study per day	Pages to study per week
K-A	89			
K-B	79			
Kindergarten total	168			

In general, kindergartners might spend 15-30 minutes a day on math, 1st-2nd graders might spend 25-40 minutes, third-fourth graders 30-60 minutes a day, and fifth-sixth graders 45-75 minutes a day. If your student finds math enjoyable, they can of course spend more time with it! However, it is not good to drag out the lessons on a regular basis, because that can then affect the student's attitude towards math.

Working space and the usage of additional paper

The curriculum generally includes working space directly on the page for students to work out the problems. However, feel free to let your students use extra paper when necessary. They can use it to draw diagrams and pictures to help organize their thoughts. Some students won't need the additional space (and may resist the thought of extra paper), while some will benefit from it. Use your discretion.

Concerning tests

While the rest of the grade levels in Math Mammoth include chapter tests, I feel that such testing is not necessary in kindergarten, as it can so easily lead to math anxiety and to viewing mathematics as a subject where correct answers are what matters most. Using tests has two sides. Even in other grade levels, the tests are optional. On the positive side, they can help the teacher know how their students are doing, and even help the student to know where they are at and to perhaps get some useful feedback. But all too often, testing leads to math anxiety. Also, students tend to study for the test only, just to forget the information right afterwards.

You can learn to assess your student(s) over the course of the year by observing them. This is even easier when you will be doing many activities with them that teach math concepts. You will be able to see where your child struggles and where they excel. You can even base a grade solely on your observations (if you need to give a grade). Or, you can use the chapter reviews as part of the grade.

The curriculum does include an optional end-of-year test, which is best used as a diagnostic or assessment test. If you do use it, it can tell you how well the child has mastered essential kindergarten math content.

Concerning the mixed reviews and the worksheet maker

The books contain mixed review lessons which practice concepts and skills from earlier chapters. They allow you to spot areas that the student has not grasped well or has forgotten. When you find such a topic or concept, you have several options:

1. Check if the worksheet maker lets you make worksheets for that topic.
   https://www.mathmammoth.com/private/Make_extra_worksheets_K.htm
2. Check for any games and resources in the Introduction part of the particular chapter in which this topic or concept was taught.
3. If you have the digital version, you can reprint a corresponding lesson from the worktext for the student.
4. Check if our online practice area at https://www.mathmammoth.com/practice/ has something for that topic.
5. Khan Academy has free online exercises, articles, and videos for most any math topic imaginable.

Frequently asked questions and contacting us

If you have more questions, please first check the FAQ at https://www.mathmammoth.com/faq-lightblue

If the FAQ does not cover your question, you can contact us https://www.mathmammoth.com/contact

Chapter 1: Counting & Writing Numbers
Introduction

Chapter 1 of *Math Mammoth Kindergarten Worktext A* focuses on counting and learning to write numbers.

Being able to count accurately is not as simple as it seems to us adults. Mastering the act of counting means that the child be able to do *all* these:

- Recite the counting sequence (one, two, three, four, ...) in the correct order and without skipping any words
- Match one counting word with exactly one item (one-to-one correspondence). The child needs to learn to pair each object with one and only one number name, and each number name with one and only one object.
- Understand cardinality: that the last word spoken tells us the total number of items in the group. In other words, children need to connect the verbal count ("1, 2, 3, 4") to the thought of "how many".

  You can test for this by having a child count some objects, and after they have the count, rearrange the *same* objects and ask again how many there are. If the child has to count again, they don't have the concept of cardinality yet.

Some other goals for this chapter are:

- Connect number words to written numerals.
- Understand that the number of objects is the same regardless of their arrangement or the order in which they were counted. Again, this is the thought of cardinality. To help a child learn it, ask them to count a number of objects twice, but starting from a different object, and help them realize the count is the same.
- Learn that each successive number name refers to a quantity that is *one* larger.
- Learn to count forward beginning from a given number, instead of having to begin at 1.

Counting is foundational to learning addition and subtraction. Many children already learn to count before kindergarten, so feel free to go through this chapter quickly if that is the case.

The worksheet maker at https://www.mathmammoth.com/private/Make_extra_worksheets_K.htm provides worksheets for writing numbers and for counting, should your student(s) need extra practice with those.

Students do more counting in chapter 6, where they will learn to count to 100 by ones and by tens. They will learn that it is quicker to count a large number of items if they are grouped into tens first. This prepares them to learn place value with two-digit numbers in first grade.

Good Mathematical Practices

- Here, and in all further studies of mathematics, children need to learn to be accurate and precise in their communication about math. Counting is a topic where it is easy for children to see the need to be accurate. We want an accurate count! True, there is a place for estimation also, but in kindergarten, the emphasis is on counting small quantities, and to learn to count accurately. Understanding one-to-one correspondence is naturally the key to that.

Pacing Suggestion for Chapter 1

Each lesson is intended to take one day. Feel free to take a day at times to just do activities and play games, without any written work. We want children to enjoy doing math, and not feel pressured.

The Lessons in Chapter 1	page	span	suggested pacing	your pacing
Counting and Sorting Objects	18	*1 page*	1 day	
Learning to Write 1	19	*1 page*	1 day	
Learning to Write 2	20	*2 pages*	1 day	
Learning to Write 3	22	*2 pages*	1 day	
Learning to Write 4	24	*2 pages*	1 day	
Practice 1	26	*1 page*	1 day	
Learning to Write 5	27	*2 pages*	1 day	
Learning to Write 6	29	*2 pages*	1 day	
How Many?	31	*1 page*	1 day	
Learning to Write 7	32	*1 page*	1 day	
Practice 2	34	*1 page*	1 day	
Learning to Write 8	35	*2 pages*	1 day	
Learning to Write 9	37	*2 pages*	1 day	
Cuisenaire Rods (optional)	39	*1 page*	1 day	
Learning to Write 10	40	*1 page*	1 day	
Counting Practice	41	*1 page*	1 day	
Practice 3	42	*1 page*	1 day	
Practice with Patterns	43	*1 page*	1 day	
More Counting	44	*1 page*	1 day	
Chapter 1 Review	45	*1 page*	1 day	
TOTALS		*26 pages*	19 days	
with optional content		*(27 pages)*	(20 days)	

Games and Activities

Board games *(Counting)*

You need: Any board game you might have where you move pieces by rolling one or two dice works to practice counting and the counting sequence.

Secret Bag *(The counting sequence, counting, estimating)*

You need: A small non-transparent bag. Small items, such as rocks, ping pong balls, spoons, buttons, etc.

Activity: Place some (up to 10) of the items in the bag. Ask the child or students to guess how many items are in the bag. You may let them even feel the bag. Once the guesses are in, pull the items out one by one (or have a child do that), with everyone counting aloud. Then repeat, taking turns.

Bean Bag Toss *(Number recognition; the counting sequence)*

You need: Bean bags, 8-10 sheets of paper, and a marker pen. Write the numbers from 1 to 10 on the papers (one number per sheet), to be the targets. If the child knows how to write numbers, they could write them.

Activity: Place the sheets of paper in order on the floor or on stairs, etc. The child will toss a bean bag on number 1, saying aloud, "One!", then on number 2, saying aloud "Two!", and so on.

Variation: Once the above activity feels easy, place the targets on the floor in a scattered order.

Animal Rescue *(Counting)*

You need: Two six-sided dice, or one ten-sided dice. About two dozen stuffed animals, plastic animals, lego figurines, or similar things. In a pinch, use beans or marbles, and ask the child(ren) to imagine they are some type of animals. An environment for the animals where they are "in danger." For example, this could be a blanket that represents seawater, or it could be something representing a barn on fire, etc.

Game play: Decide whether you will use one or two dice (or a six-sided vs. ten-sided dice). On your turn, roll the die, and "rescue" as many animals as the dice shows. Those animals become yours. In the end of the game, the player with the most rescued animals wins.

Variation: Use two ten-sided dice, to practice counting up to 20.

Tour Around the House (or Room) *(Counting)*

This game practices counting and will appeal to children who like movement.

You need: A six-sided dice. One person who stays outside the game and will bring the dice to each player to throw, wherever they are located in the room/house. Prepare a "game board" or route inside your house or classroom using pieces of cardboard, towels or similar items as game squares. The child should be able to jump from one square to the next easily.

Game play: First, each player rolls the dice one time, and the player with the highest roll starts the game. During your turn, roll the dice (the extra person will need to bring the dice to you), and jump the number of squares the dice shows. If two players land on the same space, the one that was there earlier will need to go backward one space. The first person to reach the finish wins the game.

Variation: Use two dice instead of one. The "game board" will need to be longer though!

Collect Jewels Game *(Counting)*

You need: A game piece for each player, one six-sided dice, and a pile of items for "jewels". These items can be: rocks, marbles, beans, buttons, anything you have on hand (you will need about 50 or more). You will also need to download and print the Collect Jewels game board, found here:

https://www.mathmammoth.com/download/CollectJewels.pdf

Game Play: First, each player rolls the dice one time, and the player with the highest roll starts the game.

During your turn, roll the dice, and move your piece the number of spaces the dice shows. The number on the square where you land indicates how many jewels you get from the common pile (the "bank"). The child should count these jewels on their own — after all, the purpose of the game is to practice counting.

Once every player has reached the end, the player with the most jewels is the winner. At the end is a good opportunity to practice counting larger quantities. Teach the child to arrange their pile of jewels into piles of ten, and some left over. Once done, you can help the child count by tens and add the "leftovers" to find the total, or if this is too challenging, do it yourself.

Variation: Simple counting may be too easy to some players. Players that already know addition and/or subtraction can be asked to make an addition (or subtraction) sentence with the answer of the square where they land. For example, if they land on "7", the task is to give a sum such as $3 + 4 = 7$, and after that, the player may count out their jewels out of the common pile.

Count the Bears *(Counting)*

You need: A bunch of small items to count, such as plastic counting bears, Lego figurines, other small toys, buttons, or similar items. Dominoes — about 6 to 10 times as many dominoes as there are players. (For example, for 2 players, have 12 to 20 dominoes.) Limit the dominoes to such that the sum of the dots is 10 or less, or possibly to higher numbers according to the child's counting ability.

Game play: Place the items to count in a bag or bowl in the middle. Place the dominoes face down around the bag or bowl. On your turn, take one domino and figure out how many dots it has in total. Then take that many counting items from the bag to yourself.

Encourage the child to not always start counting the dots on the domino from 1. For example, if the domino has 3 and 4 dots, if the child can see immediately that one side has 3, then they can *count on* from 3 to figure out the total: 4, 5, 6, 7.

After all the dominoes have been used up, the player with the most counting items is the winner.

Dominoes *(Counting)*

You need: A set of dominoes. It can be a smaller or a bigger set and it won't matter if some are missing. Limit the dominoes to such that the child or children are able to count the dots on them.

Activity: Just free play with dominoes is sufficient to practice counting and learning to recognize how many dots are in the dot-patterns (subitizing). The child should arrange the dominoes in the usual way, so that the end of one domino matches one end of the next one.

Simplified domino game: Place all the dominoes face down in the middle of the table (like a pond). Each player will choose five dominoes for their own "hand". The youngest player will start the game with any domino from their five, placing it in the middle of the table. At their turn, a player will place one of their dominos next to the first domino so that the ends match. If they cannot go, they draw one domino from the pond and their turn is over. Each double domino is placed sideways, perpendicular to the regular train of dominoes, and from that point, a new branch can be started.

At your judgment, you can also try to play a traditional domino game.

Matching game *(Recognizing dot patterns, counting, subitizing, number recognition)*

You need: A set of cards for this game, which you can download here. It is also provided as a PDF file in the digital version of the curriculum.

https://www.mathmammoth.com/download/Numbers-dots-matching-game.pdf

Activity: Choose a certain number of matching cards before the game. For example, you could choose a total of 16 cards, with 8 pairs. (A pair can be formed by a dot pattern matching a 10-frame card, a number matching a 10-frame, or a number matching a dot pattern.)

You can either place those cards face down in an array, for a traditional memory game, or place them face up, for a simple matching game.

The child can play alone or players can take turns finding a match.

Further Resources on the Internet

These resources match the topics in this chapter, and offer online practice, online games (occasionally, printable games), and interactive illustrations of math concepts. We heartily recommend you take a look. Many people love using these resources to supplement the bookwork, to illustrate a concept better, and for some fun. Enjoy!

https://links.mathmammoth.com/Kch1

Scan me

Counting and Sorting Objects

Choose a counting game to play from the chapter introduction. You can use it first, last, or in between the other exercises, depending on what works best for your student(s).

1. Point and count.

2. Color the stars, using a different color for each kind of star. Draw a line from each small star to the matching big star.

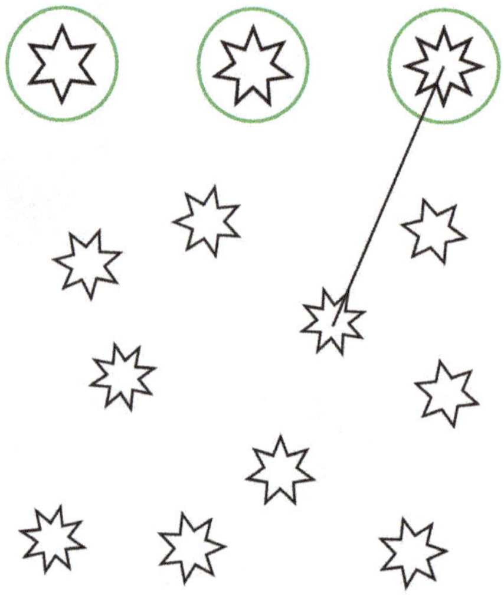

3. Write an X for each thing in the other box.

a.

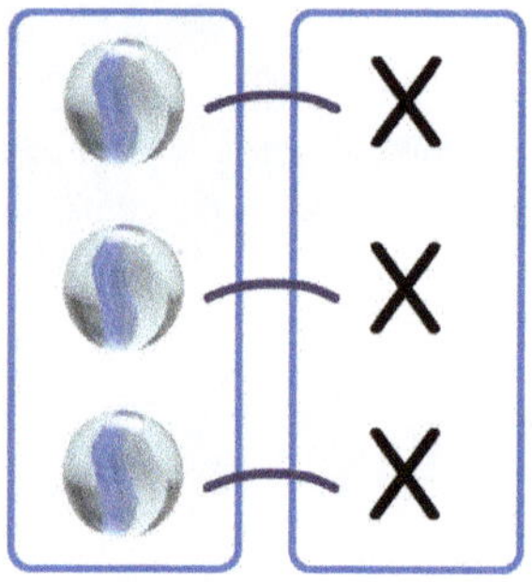

b.

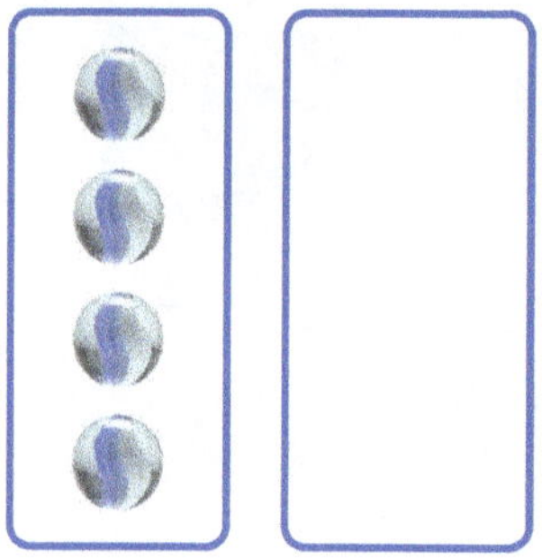

c.

4. Draw a line to match the same amounts.

Learning to Write 1

1. Trace the dots to complete the squares. Then color the squares, continuing the pattern.

2. Practice writing number 1.

3. Draw the same amount of circles below.

a.

b.

c.

d.

4. Circle the object that doesn't belong in each group.

a.

b.

Learning to Write 2

1. Practice tracing two, using colored pencils. The number can be traced multiple times.

2. More tracing practice!

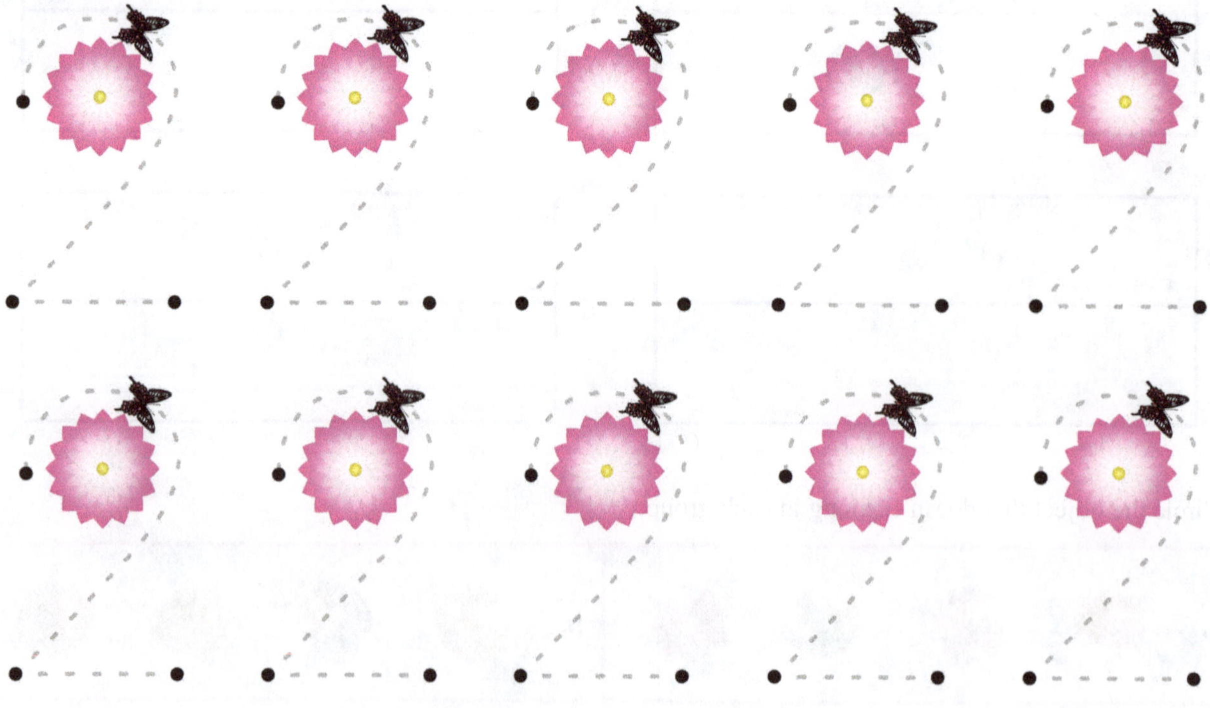

Choose a counting game to play from the chapter introduction. You can use it first, last, or in between the other exercises, depending on what works best for your student(s).

3. Practice tracing the number 2 again. Be careful! This time there are less dots!

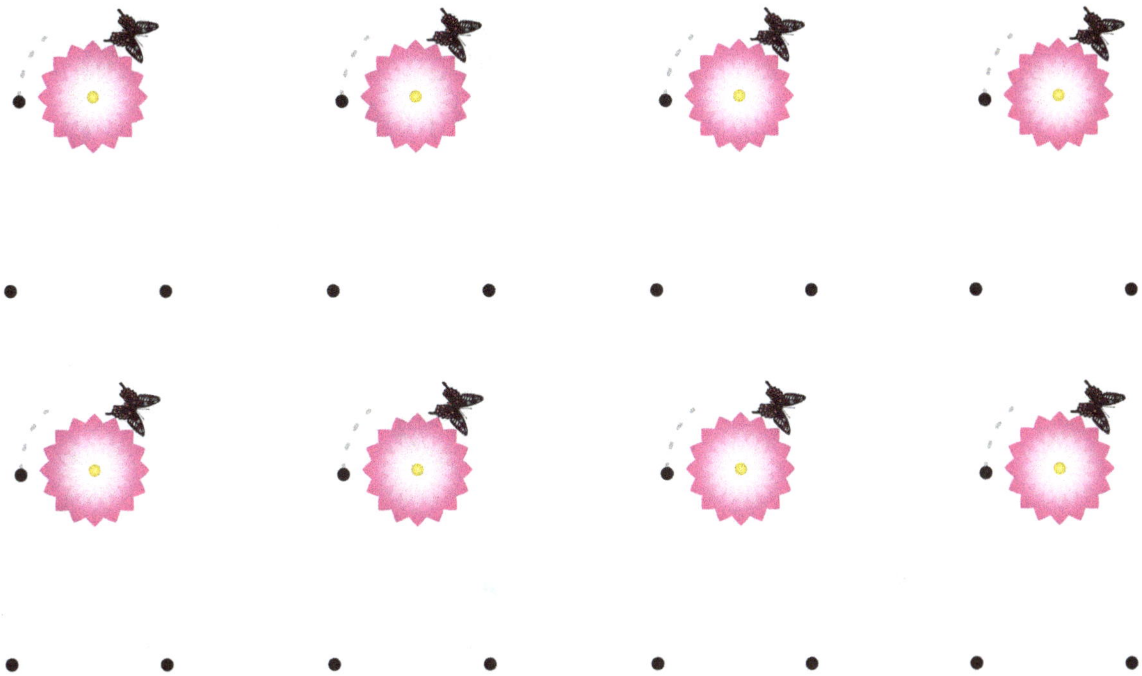

4. Practice writing the number 2 on the lines.

5. Color circles as indicated.

a. Color three circles. ○○○○○	**b.** Color five circles. ○○○○○

Learning to Write 3

1. Practice tracing three, using colored pencils. The number can be traced multiple times.

2. More tracing practice!

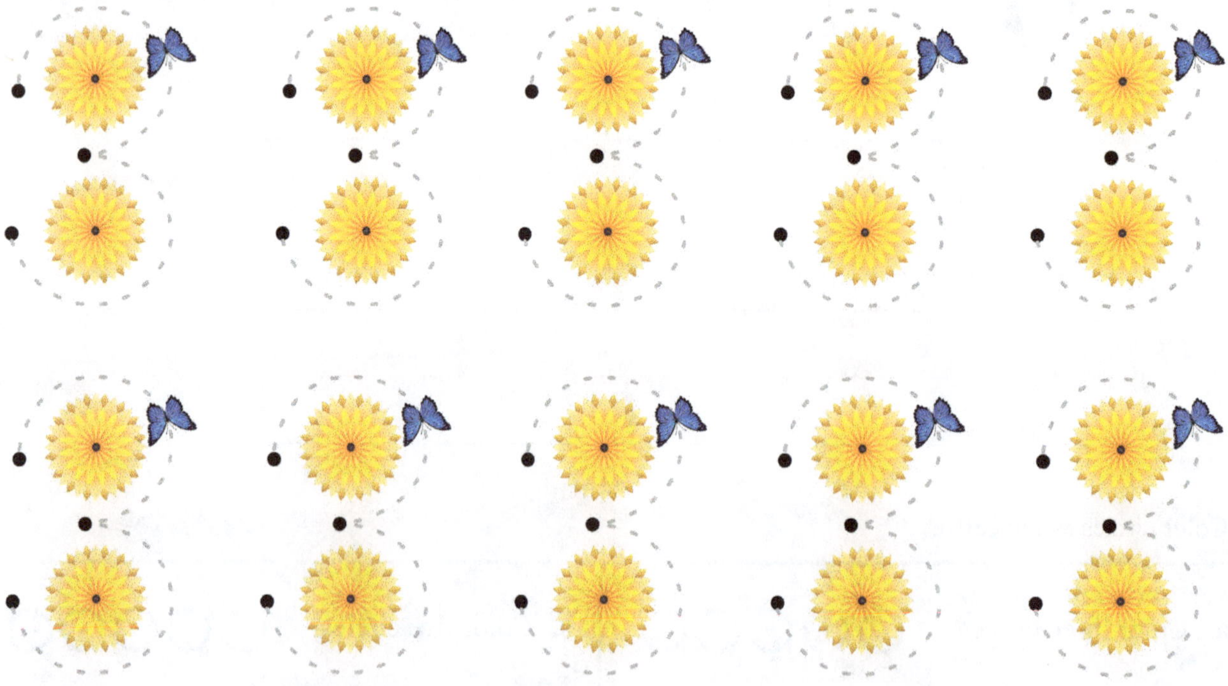

3. Write an X for each thing in the other box.

a.

b.

c.

4. Practice tracing number 3 again. Be careful! This time there are less dots!

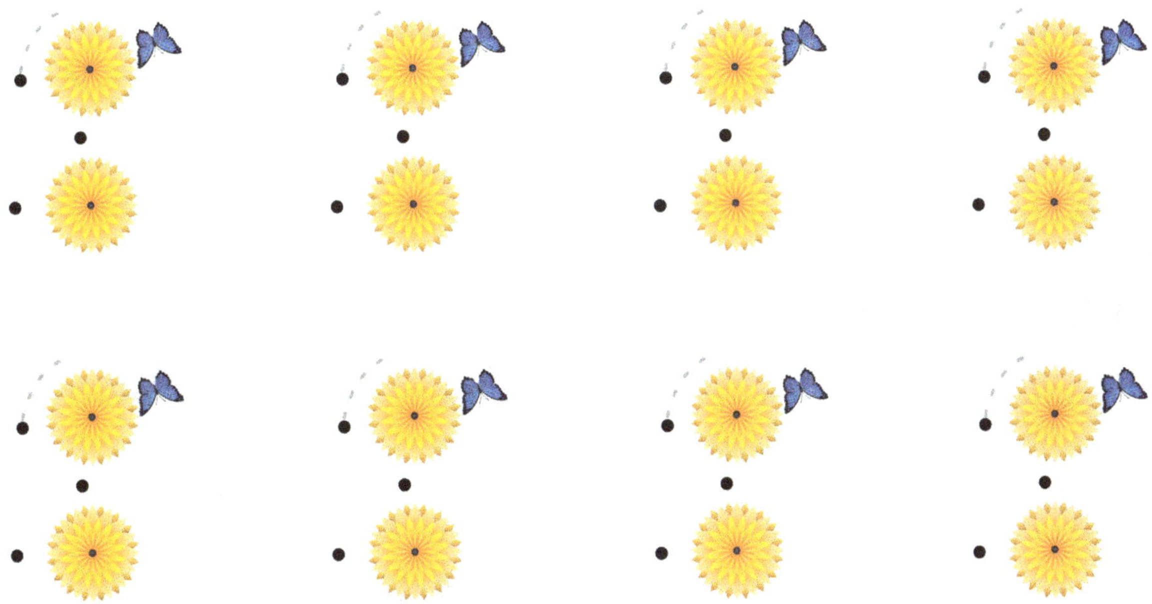

5. Practice writing the number 3 on the lines.

3

Learning to Write 4

1. Practice tracing four, using colored pencils. The number can be traced multiple times.

2. More tracing practice!

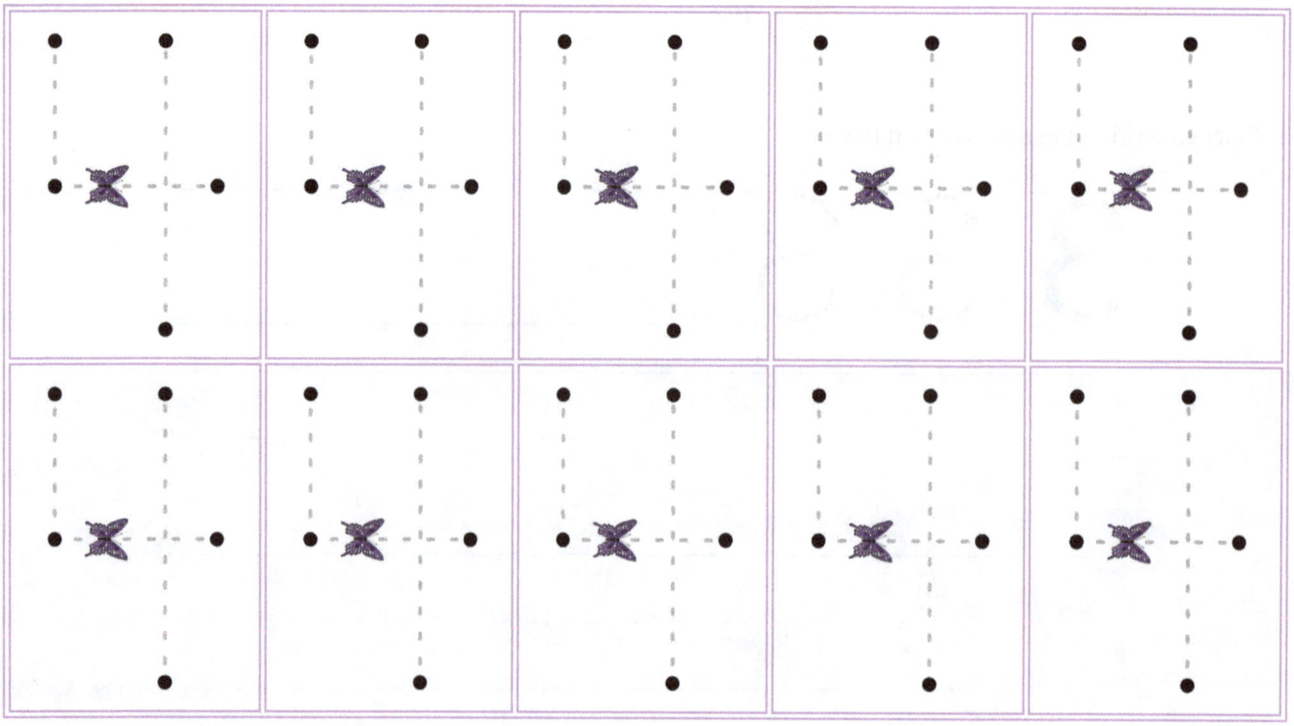

3. Practice tracing the number 4 again. Be careful! This time there are less dots!

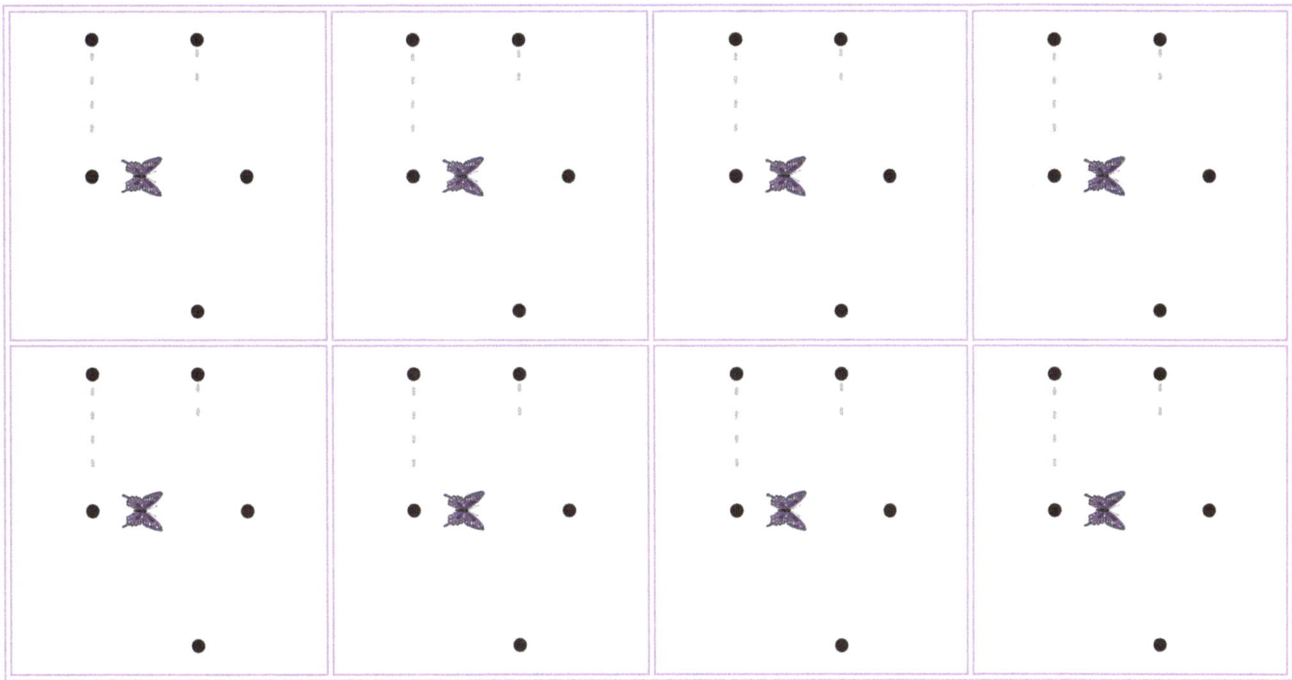

4. Practice writing the number 4 on the lines.

5. What do you notice in this picture?

What do you *wonder*?

Talk with your teacher or with other children.

Practice 1

1. Color the indicated number of turtles.

a. 1 	b. 2 

2. Write the numbers 1, 2, 3, and 4 on the lines. Then do it again.

3. There are five 2s in this picture. Find them, and color the sections they are in.

4. How many **oranges** are in each box? Count them, and write the number on the line.

a.  ______	b.  ______

Learning to Write 5

1. Practice tracing five, using colored pencils. The number can be traced multiple times.

2. More tracing practice!

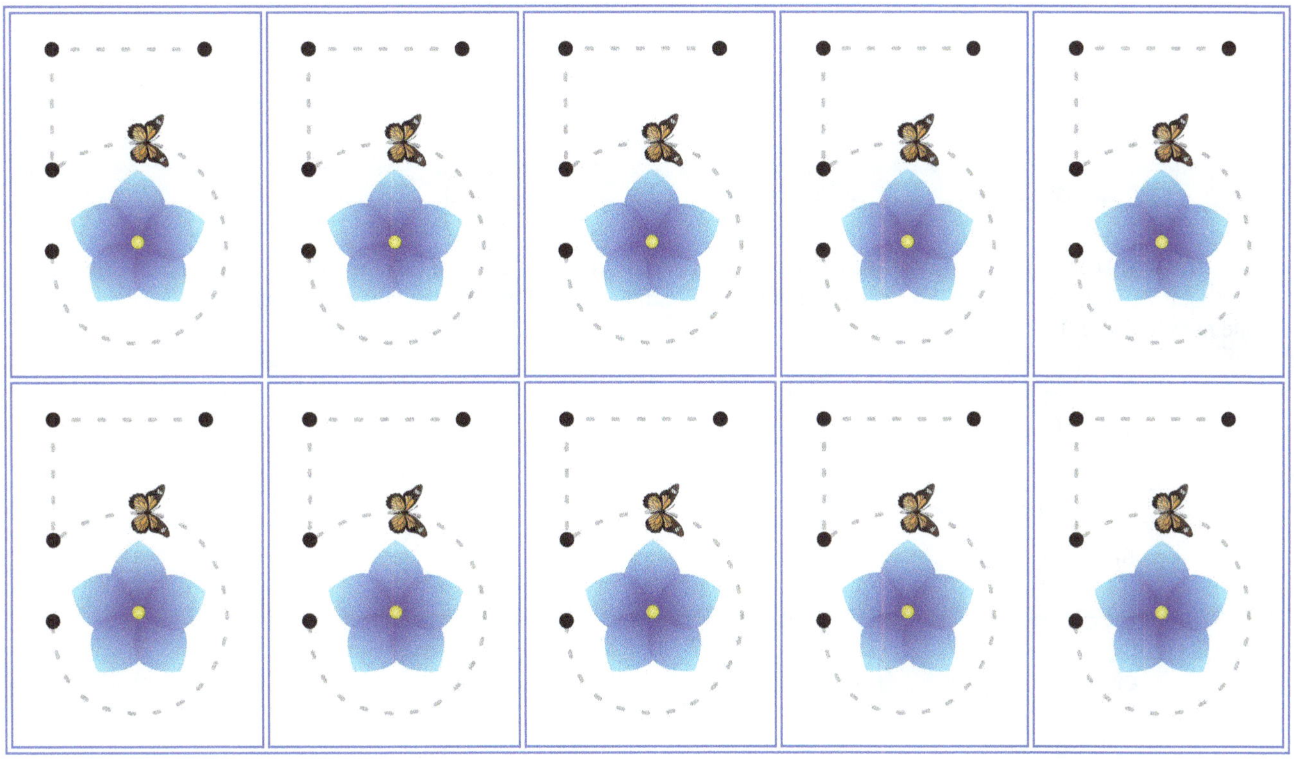

3. Practice tracing the number 5 again. Be careful! This time there are less dots!

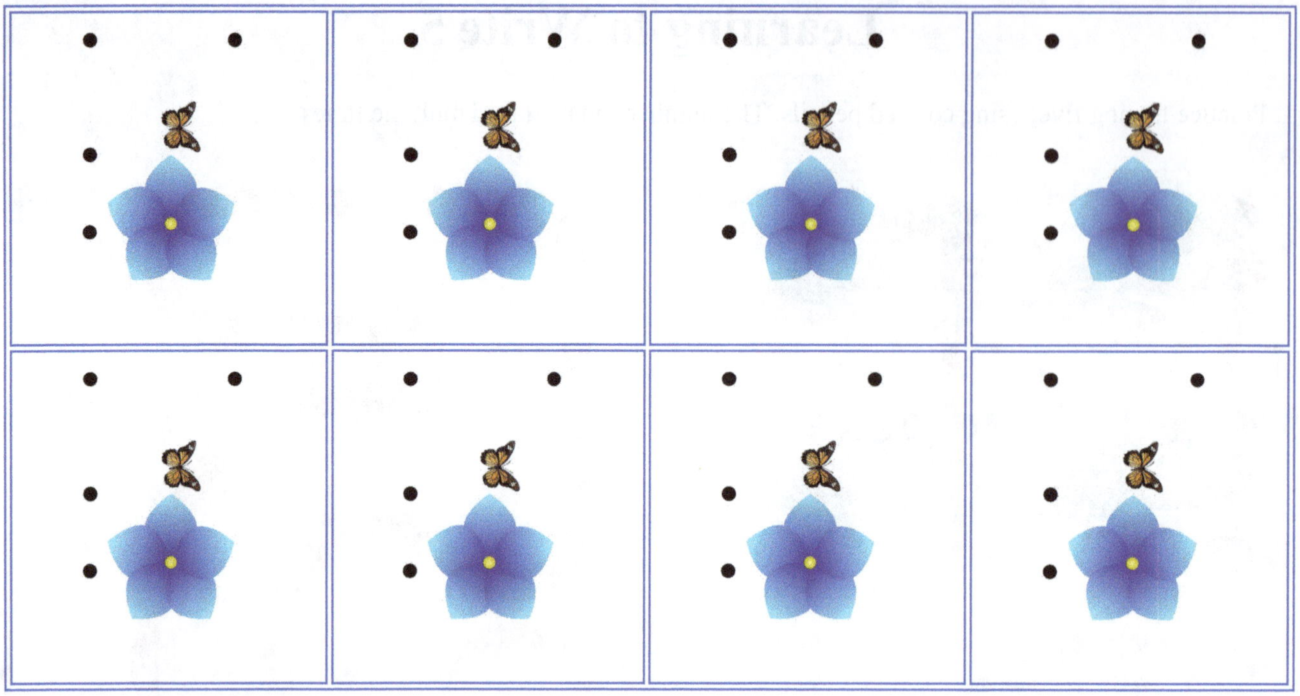

4. Practice writing the number 5 on the lines.

5. Count the circles. Then draw **one more** circle. Write the *new* total on the line. (See if the child can say the new total without counting all the circles again. **Hint:** Prompt the child by asking a question, such as "What number is one more than 3?")

a.	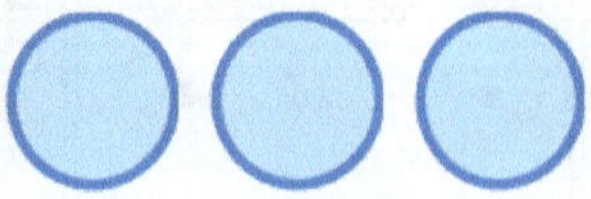	______
b.	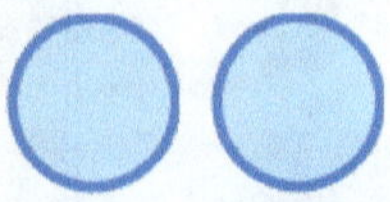	______
c.	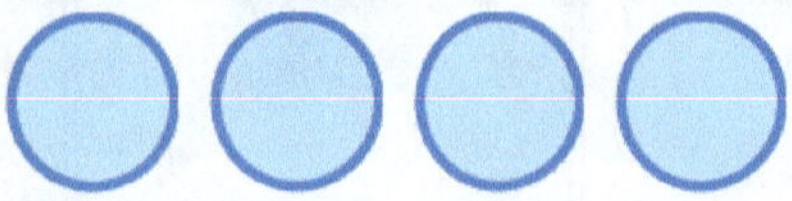	______

Learning to Write 6

1. Practice tracing six, using colored pencils. The number can be traced multiple times.

2. More tracing practice!

3. Practice tracing the number 6 again. Be careful! This time there are less dots!

4. Practice writing the number 6 on the lines.

6 6 6

5. Count the blue dots and write the number on the line.

a. ○○○○○ _____	b. ○○○○○ _____

Choose a counting game to play from the chapter introduction. You can use it first, last, or in between the other exercises, depending on what works best for your student(s).

How Many?

1. Count. Write the number on the line.

a.  3

b.  ______

c.  ______

d.  ______

2. How many? Circle the right number.

a. 

3 5 6

b. 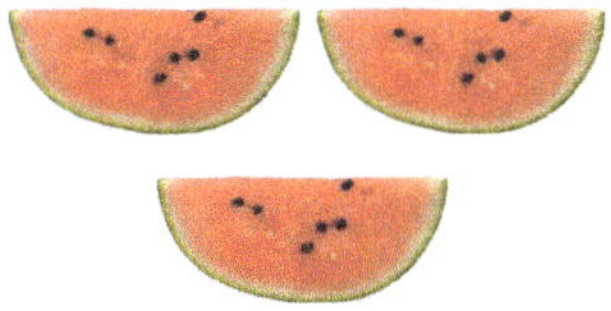

2 1 3

c. 

4 6 5

3. Count how many animals there are. Then color the animals using the color key.
For example, if there is one animal, color it red.

1 is 

2 is 

3 is 

4 is 

5 is 

6 is 

Learning to Write 7

1. Practice tracing seven, using colored pencils. The number can be traced multiple times.

2. More tracing practice!

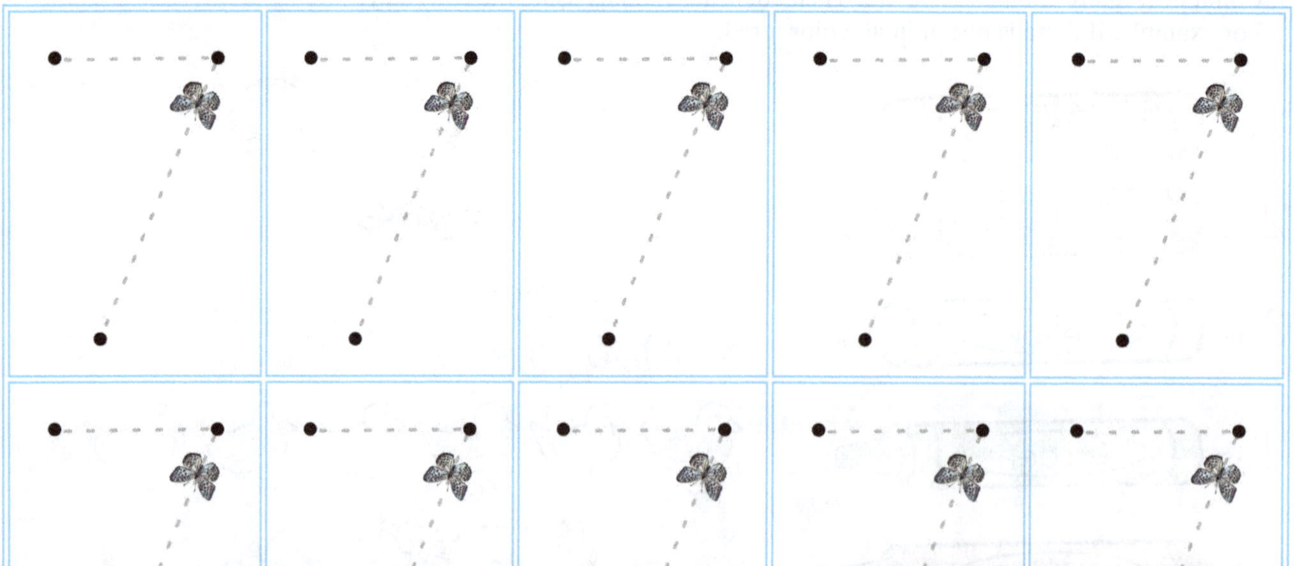

3. Practice tracing the number 7 again. Be careful! Now the dashed line is not as long.

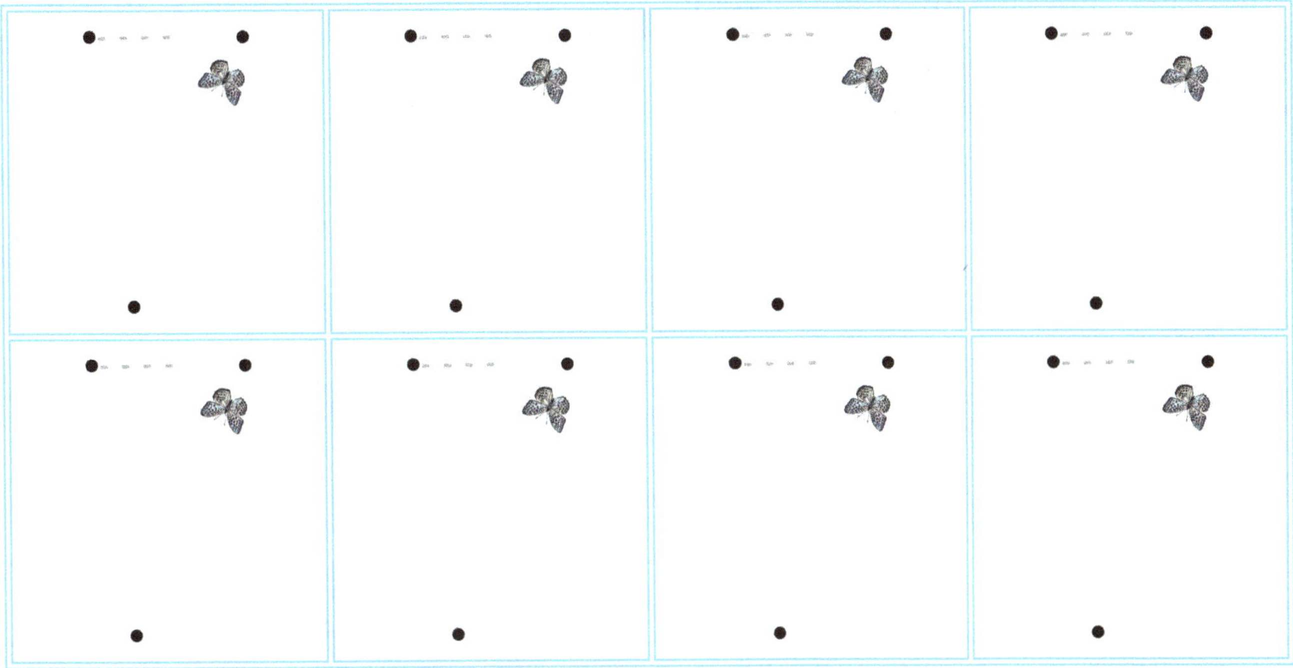

4. Write.

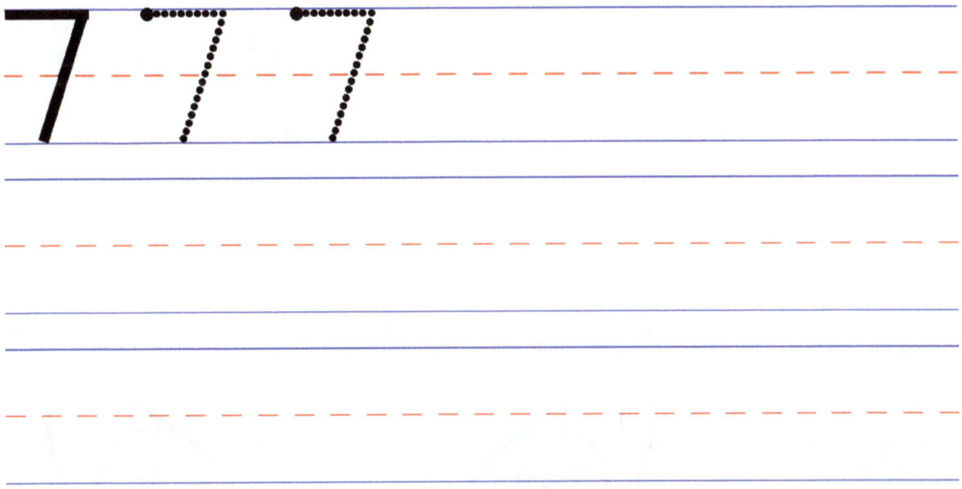

5. Count the objects. Then cross one out so there is **one less**. Write the *new* total on the line.
(See if the child can say the new total without counting all the items again. **Hint:** Prompt the child by asking a question, such as "What number is one less than 4?")

a.  ______	**b.**  ______
c.  ______	**d.**  ______

Practice 2

1. Write the numbers 5, 6, and 7 on the lines. Write each number at least three times.

2. Match.

7

3

5

4

6

3. Color the butterfly's wing according to the given color code. Then color the second wing to match the first one.

- 1 = purple
- 2 = yellow
- 3 = green
- 4 = orange
- 5 = blue

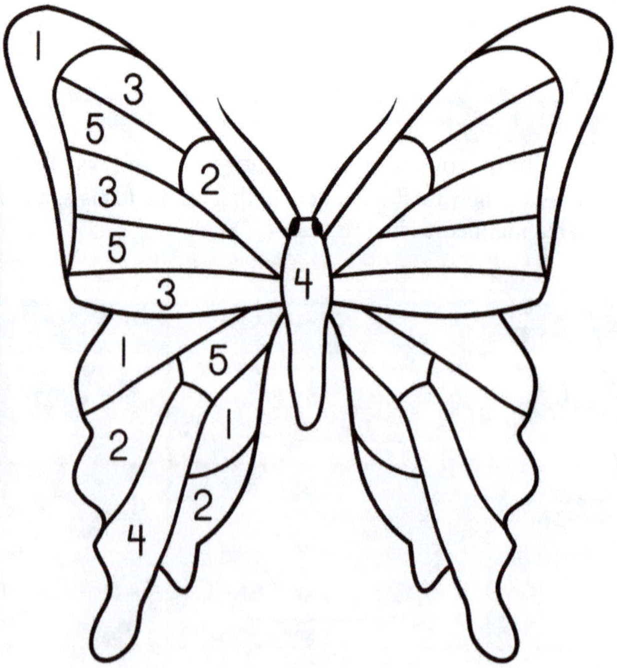

Learning to Write 8

1. Practice tracing eight, using colored pencils. The number can be traced multiple times.

2. More tracing practice!

3. Practice tracing the number again. Be careful! This time there are less dots!

4. Practice writing the number on the lines.

8 8 8

5. Cross out one of the balls so that there is **one less**. Write how many are left.

a. ________ **b.** ________ **c.** ________ **d.** ________

Learning to Write 9

1. Practice tracing nine, using colored pencils. The number can be traced multiple times.

2. More tracing practice!

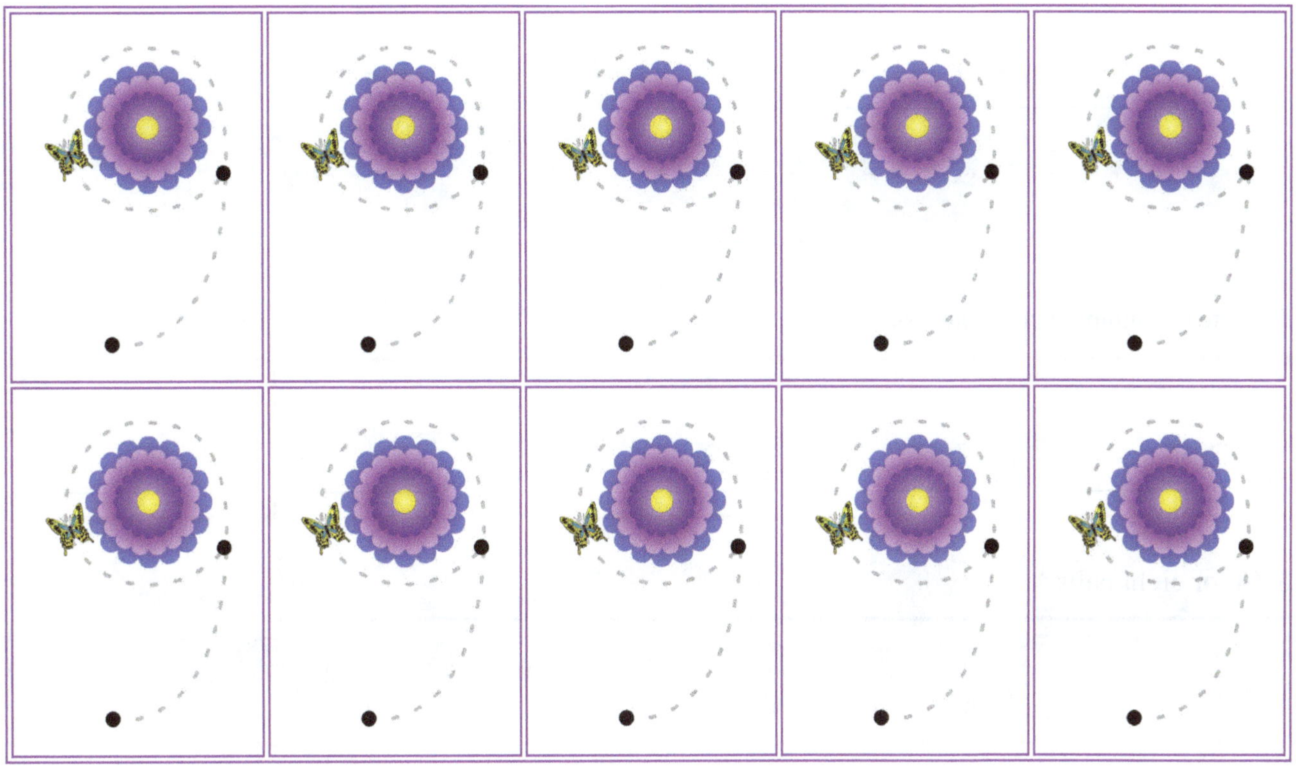

3. Practice tracing the number 9 again. Be careful! This time there are less dots!

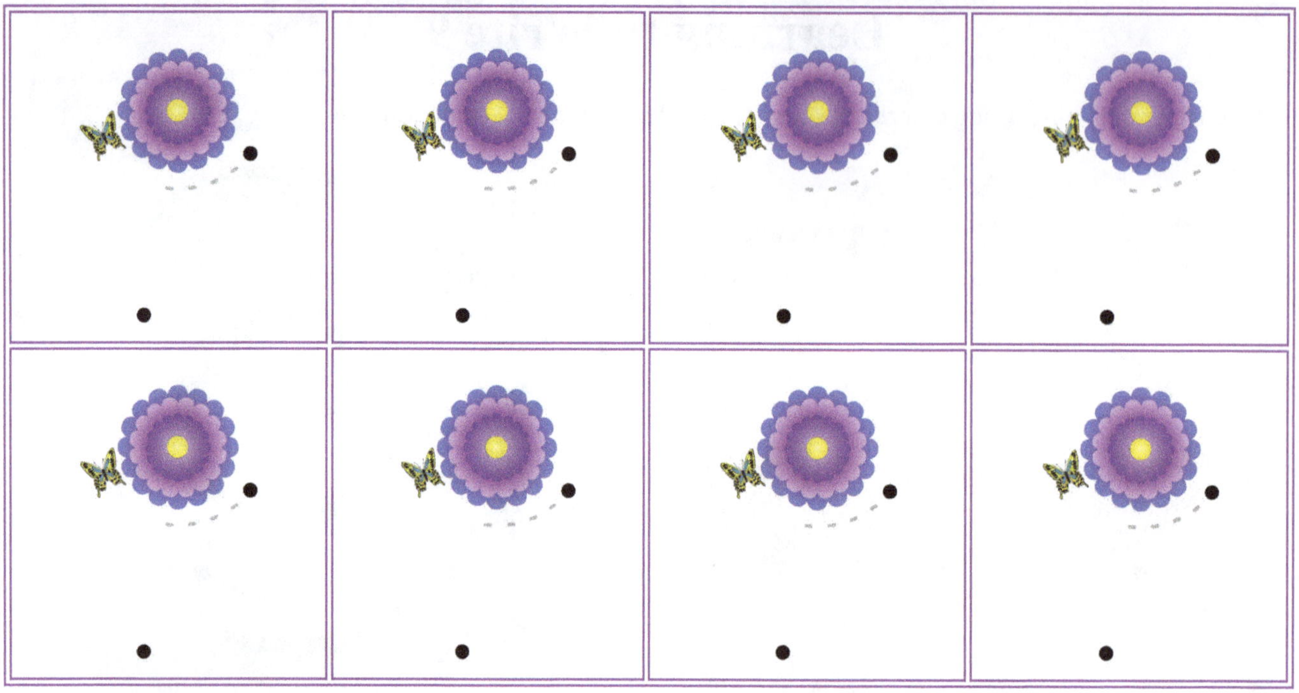

4. Practice writing the number 9 on the lines.

9 9 9

5. Write the number that is **one less**.

a. 9 ____	b. 7 ____	c. 3 ____	d. 5 ____	e. 6 ____

6. Color eight balls.

8 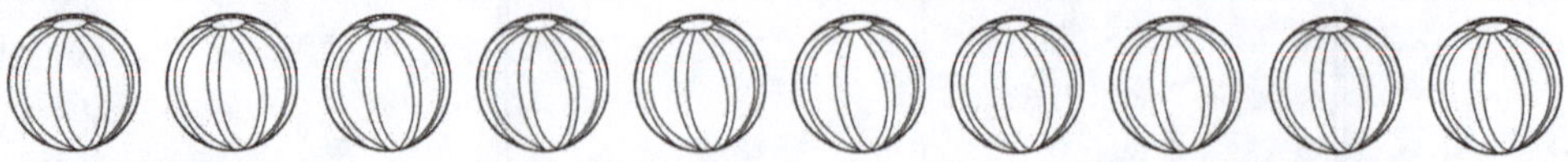

Cuisenaire Rods

(This lesson is optional.)

Cuisenaire rods are a fairly useful manipulative, and can be used to illustrate and study several kindergarten math topics. This lesson focuses on them. In subsequent lessons, there will be occasional activities that use the rods; however, you can get by without them, also.

A set of Cuisenaire rods has rods 1 cm long, 2 cm long, 3 cm long, and so on, all the way up till the longest rods being 10 cm long. The colors of the rods don't matter for our purposes; the ones with standard (traditional) colors work just as well as newer ones with other colors.

1. **Activity: rod lengths.** Find out the lengths of the Cuisenaire rods (in random order). Have the child or students find out how long each different Cuisenaire rod is, or in other words, what number it represents (from 1 through 10).

   They can do that by measuring the particular rod using the 1-rods as "counting units". After they recognize the 2-rod, they can also use that to help them figure out the length of the longer rods, etc.

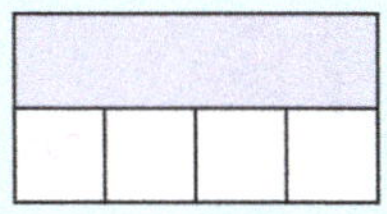

The child can find out the length of the purple rod as being 4 by placing four of the 1-rods under it.

2. **Activity/game: Guess**. Ask the child to close their eyes. Place one of the rods in their hand. They are supposed to guess how long it is, then open their eyes and check with the 1-rods or other rods if they were correct. You can also play this as a game where one player gives the next one a rod to guess, and if they get it right, they get a point.

3. **Free play.** It is important that the child is allowed to have free play time with the rods, so as to become very familiar with them.

4. **Staircase pattern.** Have the child build a staircase pattern using the rods. They can even build it both going up and down, for a symmetric pyramid.

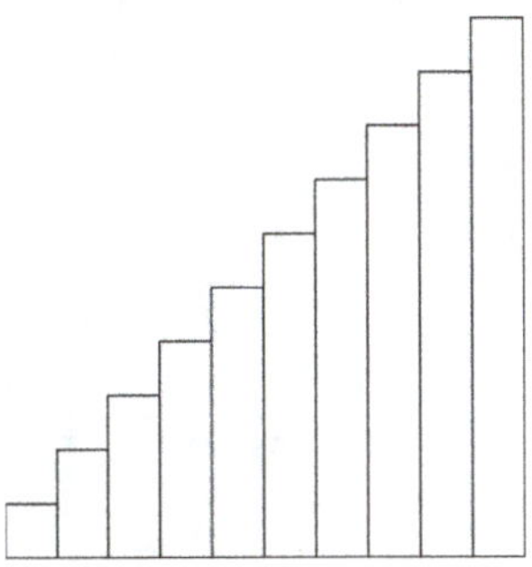

and

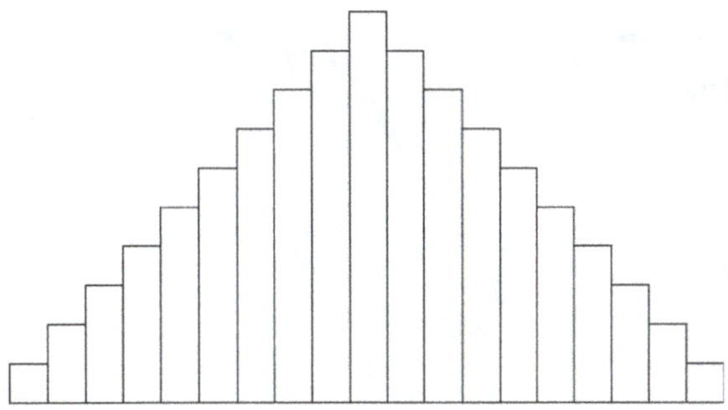

5. Now use the rods from 1-5.

   a. Find two rods that equal 3, or in other words, are together as long as the 3-rod.

   b. Find two rods that equal 4. **c.** Find two rods that equal 5.

Learning to Write 10

1. Practice writing the number 10 on the lines.

10 10 10

2. Write the number that is **one more**.

a. 6 ______	b. 9 ______	c. 2 ______	d. 7 ______	e. 4 ______

3. Color the fish using the color key.

- 6 = pink
- 7 = orange
- 8 = blue
- 9 = red
- 10 = purple

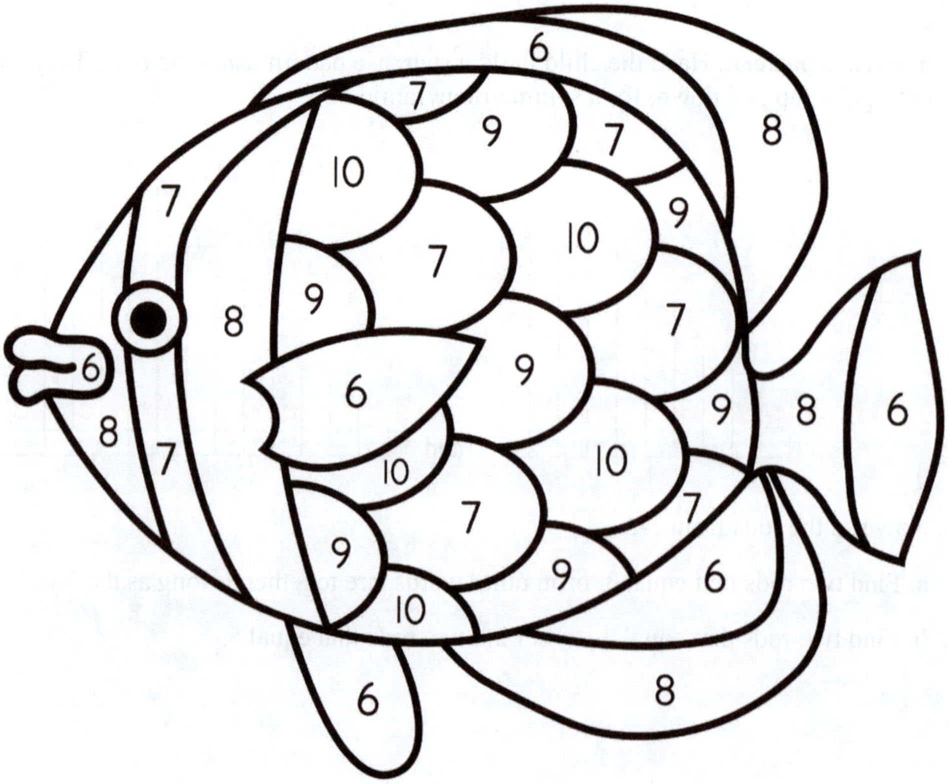

Counting Practice

1. **Activity:** Place a pile of at least 10 buttons, pencils, or other small objects in front of the child. Have the child select a specific amount of objects (between one and ten) from the pile, counting each object out loud. Return the objects to the pile, and repeat the exercise with a different amount.

Hint: In the following counting activities, the child should start counting in some logical place so that they can keep track of it: top or bottom of the array, top or bottom of the circle, etc.

2. Count. Write the number on the line.

a. ________

b. ________

c. ________

d. ________

3. Color the indicated number of objects.

a. Color six apples.

b. Color nine balls.

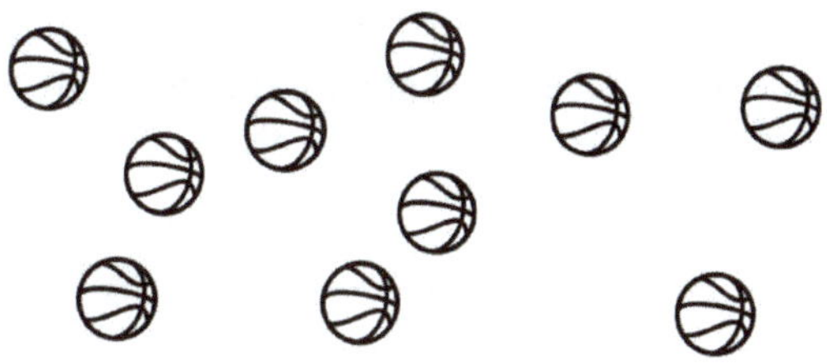

4. Count. Write the number on the line.

a.

____

b.

____

c.

____

Puzzle Corner

Find each *circle* that has a star in it.
How many are there?

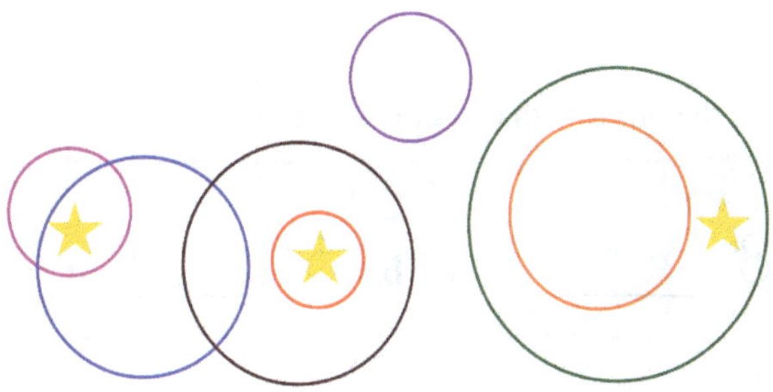

Practice 3

1. Write the numbers 1, 2, 3, and so on on the line. See if you can fit all of them from 1 to 10!

2. Count the marbles. Write the number on the line.

a.  ______	b.  ______	c.  ______

3. Count backward from five to one with your teacher! Do it several times. The pictures can help.

4. Write the numbers that are missing from the counting sequences.

a. 1, ____ , ____ , 4	c. 2 , 3 , ____ , ____ , 6
b. 7 , ____ , 9 , ____	d. 5 , ____ , ____ , 8

5. Draw a line to connect the groups that have the same number of objects.

6. Write the number that is **one more**.

a. 8 ____	b. 6 ____	c. 9 ____	d. 4 ____

Practice with Patterns

1. **Activity.** Count backward from 5 to 1 with your teacher, several times. Try it also from 7 to 1!

2. Draw shapes to continue the patterns. Optionally you can make them with Cuisenaire rods.

a. 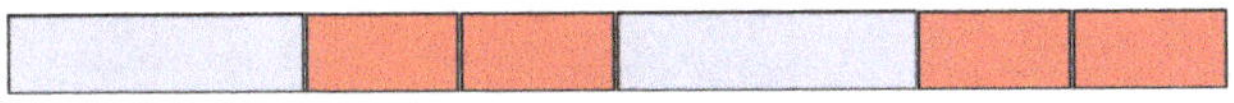

b. 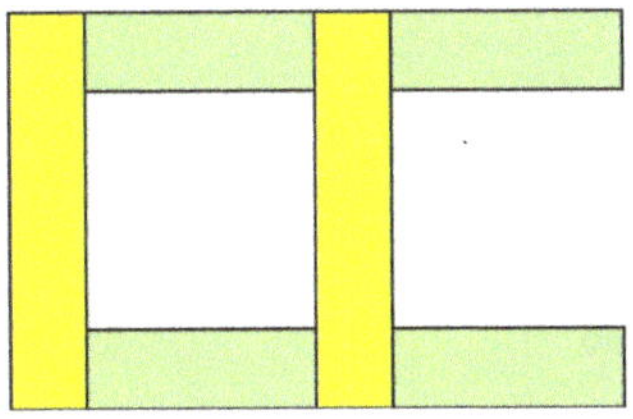

3. **Activity** (optional). Make repeating patterns with Cuisenaire rods.

4. Draw shapes to continue the patterns.

a. 

b. 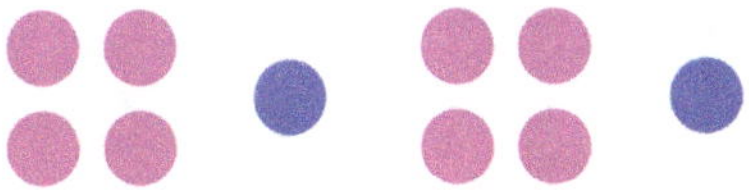

5. Continue the number patterns.

a. 1 3 1 3 ____ ____ ____ ____ ____ ____

b. 5 5 4 5 5 4 ____ ____ ____ ____ ____ ____

6. Fill in the missing numbers in this **number path**.

1	2		4			7

7. Color the shapes continuing the pattern (Y = yellow, P = purple).

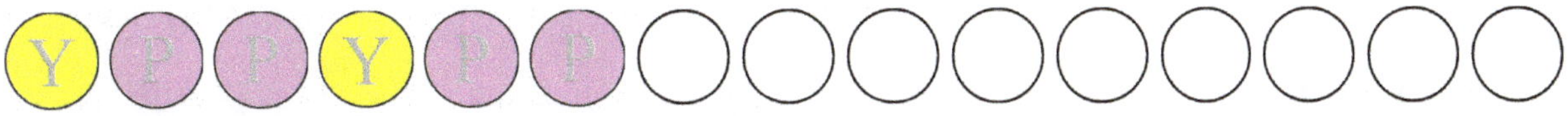

Below, make your own pattern.

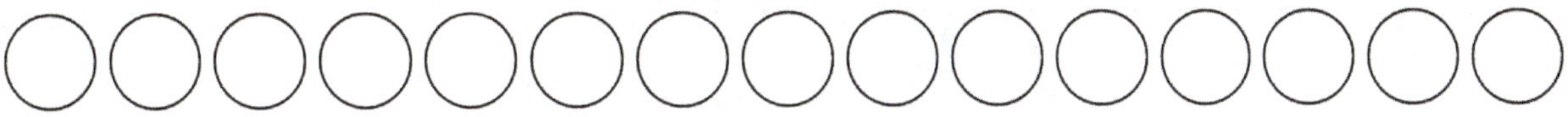

More Counting

1. **Activity.** Work in pairs. One person places 5-10 items in a scattered configuration in front of the other. The other person counts the items. Take turns, repeating the activity several times.

2. **Activity.** Count backward from five to one with your teacher or partner. Do it several times.

3. Count. Write the number on the line.

a. ________

b. ________

c. ________

d. ________

4. Fill in the missing numbers.

1		3		5		7	8

5. These number paths do not start at zero. That is alright. Fill in the missing numbers.

a.			5	6	7		9
b.				7	8	9	10

6. Color the shapes continuing the patterns (Y = yellow, G = green, B = blue).

a.	Y	B	Y	B								
b.	Y	Y	G	Y	Y	G						
c.	Y	G	B	Y	G	B						

Chapter 1 Review

1. Continue the pattern.

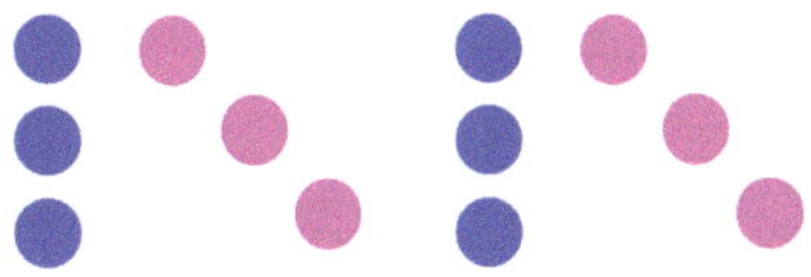

2. Write the numbers in sequence.

3 4 ______ ______ ______ ______ ______ ______

3. How many flowers and butterflies are in each box? Count them, and write the numbers on the lines.

a. 

 ______  ______

b. 

 ______  ______

4. Join the dots.
Color.

Chapter 2: Groups, Comparing, and Sorting
Introduction

In this second chapter, we focus on decomposing numbers, comparing quantities and numbers, and sorting.

Decomposing a number or quantity means separating it in parts. Children practice this skill with physical objects and visuals. It is a precursor to adding, which we will go into in the next chapter. Then in first grade, the Math Mammoth curriculum teaches the basic addition facts based on the possible decompositions of number. (For example, students will study sums of 7, which are 7 + 0, 6 + 1, 5 + 2, and 4 + 3.)

The goal here is *not* to memorize the decompositions (though if that happens, that is great), but rather, to become familiar with quantities and with the idea of breaking a number into two (smaller) parts.

Comparing two numbers (numbers that are 10 or less) is an important kindergarten skill. Start teaching this skill with an activity, using physical objects, like instructed in the lesson *More or Less? Part 1*. The goal is to go from the concrete to the abstract, where the child can tell which of the two numbers is greater or smaller without using physical objects. While the symbols < and > are introduced, there is no need to master them, nor learn to write them. That will happen in first grade, as children compare two 2-digit numbers (and in second grade, when they compare two 3-digit numbers).

Sorting objects is a simple skill that children enjoy doing. The activities in this chapter are not just about sorting but also include counting the number of objects in each category (after sorting is done), and ordering the categories by count. This prepares children for making bar graphs in first and second grades.

The lessons in the chapter often start out with an activity that uses physical objects. **These activities are important** as they are the main part of the actual teaching. Also, most children need to use physical objects to learn the concepts. It is unlikely to happen with bookwork only. So, don't skip them. They are part of the lessons, and help you as the teacher see if the child(ren) have understood the concept.

Also, the activities and games help keep math as fun and playful. We don't want math to be stressful but just the opposite! Preventing math anxiety starts here.

This introduction lists quite a few games and activities. They are referenced in the lessons, but feel free to use them as often as you wish, including sometimes taking a day just to play and do these hands-on activities.

Good Mathematical Practices

- As children find all the different ways to make 6, 7, 8, 9, and 10, they will see patterns. Mathematics has been called "the science of patterns," and thus, recognizing patterns and structures is the pure essence of mathematics.

Pacing Suggestion for Chapter 2

Each lesson is intended to take one day. Feel free to take a day at times to just do activities and play games, without any written work. We want children to enjoy doing math, and not feel pressured.

The Lessons in Chapter 2	page	span	suggested pacing	your pacing
Making Two Groups	51	*1 page*	1 day	
Making 6	52	*1 page*	1 day	
Making 7	53	*1 page*	1 day	
Making 8	54	*1 page*	1 day	
More or Less? Part 1	55	*1 page*	1 day	
More or Less? Part 2	56	*1 page*	1 day	
Making 9	57	*1 page*	1 day	

More Comparisons	58	*1 page*	1 day
Making 10	59	*1 page*	1 day
Sorting Items	60	*2 pages*	1 day
Comparison Practice	62	*1 page*	1 day
Sort Some More	63	*1 page*	1 day
Mixed Review Chapter 2	64	*1 page*	1 day
Chapter 2 Review	65	*1 page*	1 day
TOTALS		*15 pages*	14 days

Games and Activities

I Get, You Get *(Decomposing numbers)*

You need: A bunch of small items (ten is plenty).

Activity: Choose a number of items (say 7) and place them on a table. Let the child count to find out how many there are. Then grab some items (e.g. 2) and say, "If I get two, what do you get?" The child needs to then figure out how many are left for them, e.g. "I get five."

Take turns, so that at times, the child will take some of the items (e.g. 4), saying, "If I get four, then how many do you get?" And then you will take the rest and say, "I get three."

The idea is to learn the concept of decomposing a number into two parts. You can also encourage the child or students to find all the different ways to decompose the same number into two parts.

Some Went Hiding *(Decomposing numbers)*

You need: Up to 10 small objects, such as marbles, pennies, beans, or rocks.

Game play: Choose a target number, such as 7. The first player shows that many objects to the next player, and then quickly hides *some* of them behind his back (without showing how many). He shows the rest of them to the next player, who has to say how many "went hiding." If the player gives the right answer, it is then his turn to hide some and ask the next player to answer. If he gives a wrong answer, he forfeits his turn. This game appeals best to young children.

Variation: Instead of getting a turn to hide objects, the player who answers correctly may gain points or other rewards for the right answer.

Cuisenaire Rods *(Decomposing numbers)*

You need: Cuisenaire rods. These are rods with lengths of 1 cm, 2 cm, 3 cm, etc. all the way up to 10 cm, with a different color for each length.

Activity: First, allow some free play time with the rods and make sure the child knows how to determine the length of each rod (e.g. by checking how many of the 1-cm rods equals the length of the rod).

Then, take one of the rods (except the shortest ones), such as the 5, and ask the child to find two other rods that are as long as that rod. The child may find 2 and 3. Then ask the child to find some *other* combination of two rods that are equal to 5. After all the combinations are found, go to some other number, such as 6.

To get the most benefit from this activity, ask the child to also write down the rod combinations they find using numbers. For example, for 5, the child could write 2 + 3 = 5 and 1 + 4 = 5, or they could write them like this:

5	**5**
2 and 3	4 and 1

Which Is More? (or Less) *(Comparing numbers)*

This activity helps children understand the concepts of more and less by using physical items. They will compare two quantities (e.g. 6 and 8) by placing that many items side by side.

You need: A bunch of small items, such as counting bears, Lego figurines, plastic animals, or buttons. An ice cube tray that has two columns for each child or group. (Alternatively, you can draw a two-by-ten grid on paper.) A set of number cards. The greatest number in the set of number cards should be the same as the number of slots in the longer side of the ice cube tray.

Activity: Place the deck of number cards face down on the table. The child will draw one card, count out that many items, and place them in one column of the ice-cube tray. Then they will draw another card, count out another set of items, and place those in the *other* column of the ice-cube tray. Then they will determine which is more (which number is bigger).

Variation: Ask the child to find out which number is smaller (or which quantity is less).

Guess My Number! *(Comparing numbers)*

Game Play: This is a game or activity for two players; however you can also adapt it to groups. One player chooses a number in their mind (from 1 to 10) and says, "Guess my number!" The other player makes a guess, such as "Three." The first player then either admits that is correct, or states whether their number is more or less than the guess: "My number is greater than 3." The second player then guesses again, and again, the first player gives feedback, until the second player guesses the number.

To adapt the game to small groups, one person chooses a number, and all the others do the guessing.

The Largest Number *(Comparing numbers)*

You need: A set of number cards from 1 to 10.

Game Play: Place the deck of number cards face down on the table. Each player will draw one card. The player with the largest number gets to keep all the cards from all the players. If there is a tie between some players, each of those players will draw one more card from the deck to resolve the tie. The player with the largest number gets all the cards from the tie resolution plus all the cards from the round with the tie.

Play until the deck is used up. The player with the most cards in their personal pile wins. Alternatively, just play until the deck is used up without determining a winner.

Put More in the Cup! *(Comparing numbers)*

You need: A bunch of small items, such as counting bears, Lego figurines, plastic animals, or buttons. Two cups or containers for each player; one cup is bigger than the other. A set of number cards from 3-10.

Game play: Set the deck of cards face down on the table. Set the items in a container or bag. On their turn, the player first takes a card from the deck. Then they count out that number of items from the bag, placing the items in their two cups so that the bigger cup has *more* items than the smaller cup. The other players or a game leader will check their work. If they did it correctly, they get to keep the card they drew, placing it in their personal pile; otherwise the card goes back to the bottom of the deck.

Then it is the next player's turn. The game continues until all the cards in the deck are used up. The winner is the one with most cards in their personal pile.

Variation: Instead of keeping the card, the player will keep the items in the cups (placing them elsewhere). The winner is the player with most items at the end of the game.

List by Count *(Ordering counts of objects)*

This activity practices ordering three quantities (e.g. 2, 8, and 5). In kindergarten, children don't yet need to order three or more numbers from the smallest to the greatest using written numerals only, but they do need to learn to do this with physical objects.

You need: A bunch of small items, such as counting bears, Lego figurines, plastic animals, beans, or buttons. Three bowls or containers. A set of number cards from 1 to 10.

Activity: Place the deck of number cards face down on the table. The child will draw three cards. If some of the numbers are the same, they need to draw enough cards so that all three numbers are different. Then, the child will count out those numbers of objects, one quantity into each bowl. Lastly, the child will set the three bowls in order from the one with the least number of objects to the one with the most.

Variation: Have the child order four quantities instead of three.

Sort, and List by Count *(Categorizing objects, and ordering the categories by count)*

This activity practices categorizing items, and then listing the categories in order of count.

You need: A bunch of items that can be categorized into 3-5 categories, such as by color, size, or just being different objects. For example, you could use different colored paperclips or crayons; short, medium, and long pencils; three or four different kinds of beans; buttons of different sizes; small, medium, and large stuffed animals, etc. Three to five bowls, bags, or containers to hold the items.

Activity: Have the child sort the items into the bowls or containers. Lastly, have the child set the bowls in order from the one with the least number of objects to the one with the most.

Games and Activities at Math Mammoth Practice Zone

Sorting — Online Practice
Sort items by dragging them on screen to the correct box. The instructions are provided as both text and audio.
https://www.mathmammoth.com/practice/sorting-game

Beach Comparisons
Children will have fun comparing numbers and numbers of objects on the beach! Three different game modes are provided.
https://www.mathmammoth.com/practice/beach-comparisons

Further Resources on the Internet

These resources match the topics in this chapter, and offer online practice, online games (occasionally, printable games), and interactive illustrations of math concepts. We heartily recommend you take a look. Many people love using these resources to supplement the bookwork, to illustrate a concept better, and for some fun. Enjoy!

https://links.mathmammoth.com/Kch2

Scan me

Making Two Groups

1. **Activity.** Show the child 4 small items, and later 5. Show the child how the items can be separated into two groups (e.g. 1 and 3). Then ask the child to find *different* ways (as many as they can) to separate the items into two groups. Work together as necessary. You can also use Cuisenaire rods.

2. **Game.** Play "I Get, You Get" and/or "Some Went Hiding" with 4 objects, and then later with 5 objects. (See chapter introduction.)

3. Draw a line to separate the circles in two groups as indicated. Color each group with a different color.

a. 4	**b.** 4	**c.** 4
● | ● ● ●	○ ○ ○ ○	○ ○ ○ ○
1 and 3	2 and 2	3 and 1
d. 5	**e.** 5	**f.** 5
○ ○ ○ ○ ○	○ ○ ○ ○ ○	○ ○ ○ ○ ○
3 and 2	2 and 3	1 and 4

4. Separate the shapes into two groups with a line. Write how many are in the second group.

a. 4	**b.** 5	**c.** 4
△ △ △ △	△ △ △ △ △	△ △ △ △
1 and _____	2 and _____	3 and _____

5. Using a different color, draw more circles so that the total number of circles is the number indicated.

4 ● ●	3 ●	5 ● ● ● ●
3 ● ●	5 ● ● ●	4 ●

Making 6

1. **Activity** (optional). Use Cuisenaire rods. Find all the possible ways that two rods can equal 6.

2. **Game.** Play "I Get, You Get" and/or "Some Went Hiding" with 6 objects. (See chapter introduction.)

3. **Activity.** Count backward from five to one with your teacher. Do it several times.

4. Draw a line to separate the circles in two groups as indicated. Write how many are in the second group. Color each group with a different color.

a. 6 ○○○○○○ 1 and ______	**b.** 6 ○○○○○○ 2 and ______
c. 6 ○○○○○○ 4 and ______	**d.** 6 ○○○○○○ 3 and ______

5. Tell a story about this picture. Use numbers in your story.

6. Separate the dots into two groups, in different ways. Write how many are in each group.

a. 5 ●●●●● ______ and ______	**b.** 5 ●●●●● ______ and ______	**c.** 5 ●●●●● ______ and ______

Making 7

1. **Activity** (optional). Use Cuisenaire rods. Find all the possible ways that two rods can equal 7.

2. **Game.** Play "I Get, You Get" and/or "Some Went Hiding" with 7 objects. (See chapter introduction.) 

3. Draw a line to separate the circles in two groups as indicated. Write how many are in the second group. Color each group with a different color.

a. 7 ○○○○○○○ 5 and ____	**b.** 7 ○○○○○○○ 3 and ____
c. 7 ○○○○○○○ 6 and ____	**d.** 7 ○○○○○○○ 4 and ____

4. Write how many there are of each kind of object. Then write the total number of objects in the box.

a. 1 and 4 [5]

b. ____ and ____ [ ]

c. ____ and ____ [ ]

d. ____ and ____ [ ]

Making 8

1. **Activity** (optional). Use Cuisenaire rods. Find all the possible ways that two rods can equal 8.

2. **Game.** Play "I Get, You Get" and/or "Some Went Hiding" with 8 objects. (See chapter introduction.)

3. Draw a line to separate the circles in two groups as indicated. Write how many are in the second group. Color each group with a different color.

a. 8

5 and ______

b. 8

2 and ______

c. 8

7 and ______

d. 8

4 and ______

e. 8

3 and ______

f. 8

6 and ______

4. Using a different color, draw more circles so that the total number of circles is 6.

a. 6

b. 6

c. 6

5. Draw more dots of a different color to make the given number. Write how many dots there are of each color.

a. 7

______ and ______

b. 6

______ and ______

c. 4

______ and ______

More or Less? Part 1

1. **Activity.** Do the "Which is More?" activity from the chapter introduction. Use the activity to check that the child knows the ideas of "more" and "less".

2. Which is more, 8 or 6? Circle the bigger number.

8

6

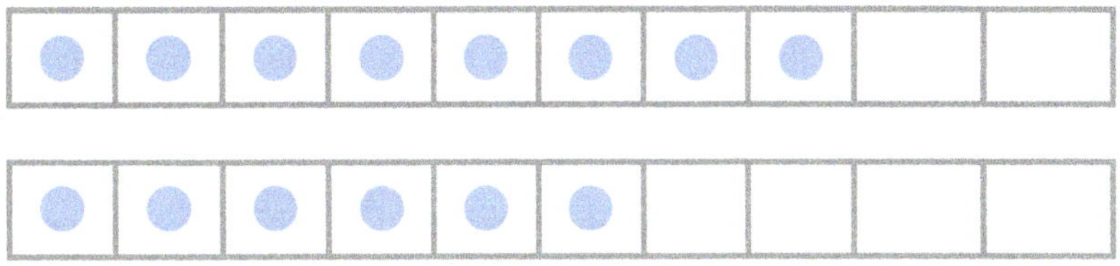

3. Now draw dots yourself to find out which is the *bigger* number. Then circle the bigger number.

a. 4 / 3

b. 5 / 7

4. Now draw dots yourself to find out which is the *smaller* number. Then circle the smaller number.

a. 5 / 6

b. 8 / 7

5. Circle the group that has **more** objects.

a.

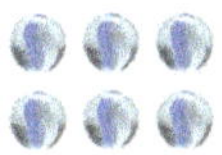

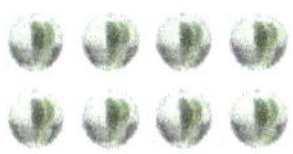

b.

6. Circle the group that has **less** objects.

a.

b.

More or Less? Part 2

1. **Activity.** (optional) If you feel your child/students need it, do the "Which is More?" activity from the chapter introduction again.

2. **Game.** Play "The Largest Number" game from the chapter introduction. If necessary, you can let the child/students use physical objects to determine which number is larger (e.g. by lining up the two groups of objects and seeing which group makes a longer line).

3. Count. Write the total on the line. Then circle the bigger number.

 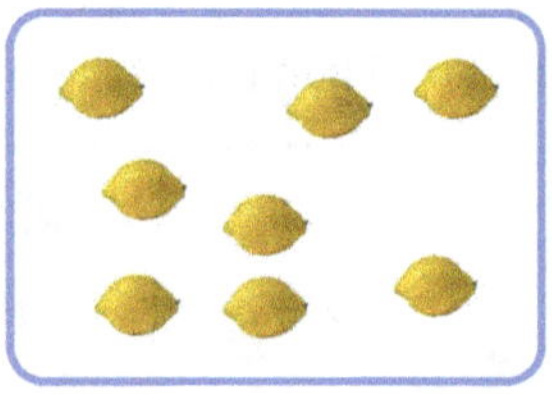

a. _______ _______

b. _______ _______

4. Count each group of objects and write the total on the line. Then circle the number that is LESS.

a. _______ _______

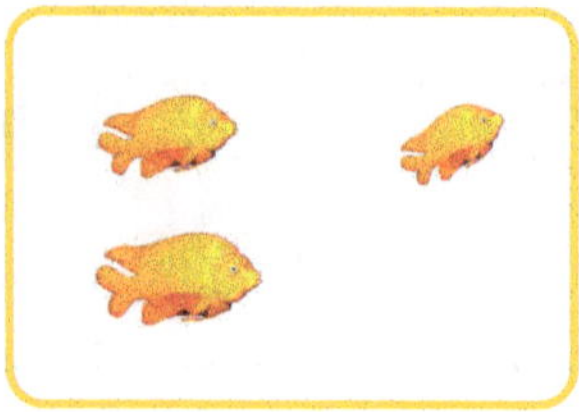 

b. _______ _______

5. Circle the bigger number. If you want to, you can draw dots to help you.

a. 4   8

b. 10   9

c. 6   7

Making 9

1. **Activity.** Count backward from five to one with your teacher. Do it several times.

   Then count backward from 10 to 6. Do it several times also.

2. Draw a line to separate nine dots in two groups, in **different ways**. Write how many are in the first and in the second group. Color each group with a different color.

a. 9 ○ ○ ○ ○ ○ ○ ○ ○ ○ _____ and _____	**b.** 9 ○ ○ ○ ○ ○ ○ ○ ○ ○ _____ and _____
c. 9 ○ ○ ○ ○ ○ ○ ○ ○ ○ _____ and _____	**d.** 9 ○ ○ ○ ○ ○ ○ ○ ○ ○ _____ and _____
e. 9 ○ ○ ○ ○ ○ ○ ○ ○ ○ _____ and _____	**f.** 9 ○ ○ ○ ○ ○ ○ ○ ○ ○ _____ and _____

3. Write how many there are of each kind of flower. Then write the total number of flowers in the box.

a. ____ and ____ ☐

b. ____ and ____ ☐

c. ____ and ____ ☐

d. ____ and ____ ☐

More Comparisons

The symbol $>$ means "**is greater than**". The symbol $<$ means "**is less than**".

To help the child learn the symbols, compare them to a "hungry alligator's mouth." The mouth always opens towards the **bigger** number.

After the child has completed this page, read each comparison aloud, having them repeat what you say. For example, "Seven is greater than four."

The important thing is to be able to tell which number is greater.
We will practice the symbols more in first grade.

6  9

"6 is less than 9."

1. Count each group of dots and write the total in the box. Circle the number that is **greater** (more).

a. 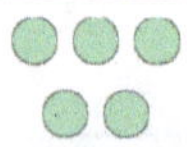 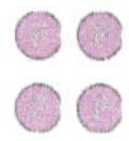

□ $>$ □

b. 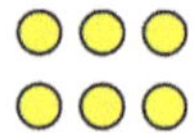 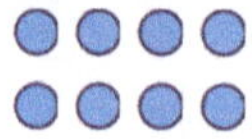

□ $<$ □

2. Circle the number that is **greater** (more).

$7 > 4$	$2 < 3$	$6 > 5$	$10 > 9$
$3 < 5$	$9 > 7$	$8 < 10$	$6 < 8$

3. Count each group of objects and write the total in the box. Then circle the smaller number.

a.  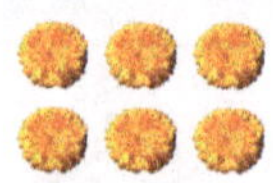

□ $<$ □

b.  

□ $>$ □

4. Circle the smaller number.

$3 > 1$	$4 < 6$	$8 < 10$	$6 > 1$
$9 > 5$	$7 < 9$	$10 > 9$	$5 < 7$

Making 10

1. **Activity.** Use your 10 fingers to show two groups. Think of the fingers that are up as one group, and the fingers that stay down as the other. One way is this: What other ways are there?

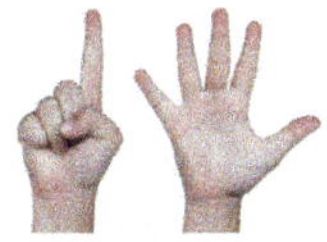

(4 fingers down, 6 up)

2. Draw more dots of a different color so there is a total of 10 dots.
Write how many dots there are of each color.

a. 10

______ and ______

b. 10

______ and ______

c. 10

______ and ______

d. 10

______ and ______

e. 10

______ and ______

f. 10

______ and ______

3. Circle the group that has more things. Then count ALL (both groups). Write the total in the box.

a. ☐

b. ☐

c. ☐

4. Match groups of lemons so that the **total** number of lemons is 10.

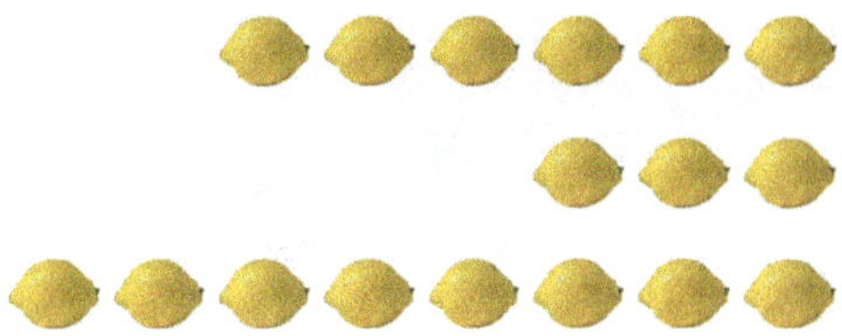

Sorting Items

1. **Activity:** Choose varying quantities of objects (less than or equal to 10) that can be classified by some attribute, such as by color, size, or type of object. For example, you could have 3 or 4 different kinds of beans to sort, or a mix of objects such as buttons, paperclips, beans, and rocks.

   Have the child sort the objects into piles and count the objects in each pile. Then, have the child sort the piles by count; in other words, place the piles in order from the smallest amount of objects to the largest amount of objects.

Note: For the following activities, if the child doesn't yet know how to write, you can prompt them to tell you the answers and write them down for the child.

2. **a.** Count how many circles there are of each **color**. Record the counts below.

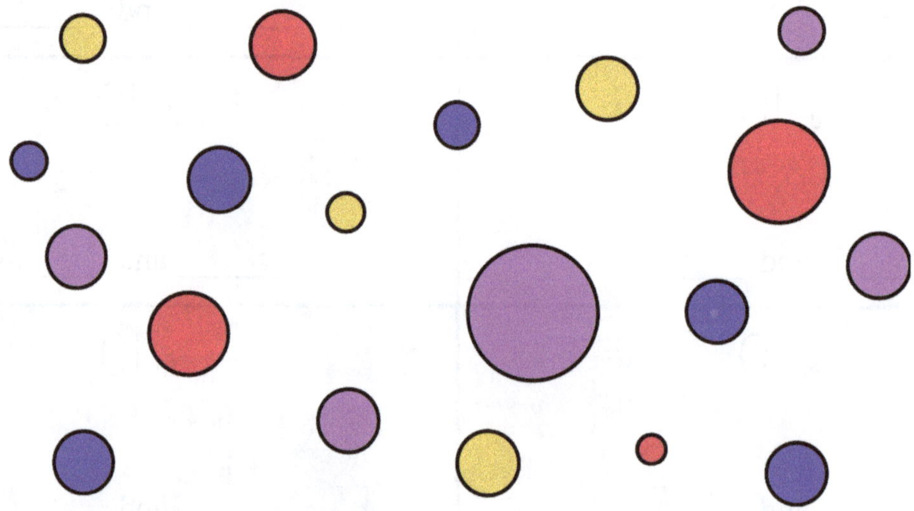

______ blue circles ______ red circles ______ yellow circles ______ purple circles

b. Now list the counts of the different types of circles, from the least count to the greatest.

______ ______________________ circles

______ ______________________ circles

______ ______________________ circles

______ ______________________ circles

3. **a.** Count how many there are of each **kind of animal**. Record the counts below.

______ snails ______ birds ______ butterflies ______ frogs

b. Now list the counts of each kind of animal in order, from the least count to the greatest.

______ ____________________________

______ ____________________________

______ ____________________________

______ ____________________________

Comparison Practice

1. **Game.** Play "Guess My Number" from the chapter introduction.

2. Draw enough dots in the first box so that there is a **greater** amount than in the second box.

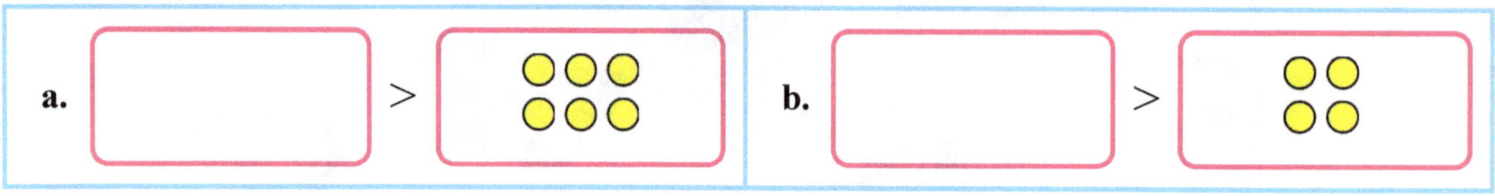

3. Circle the greater number. Practice saying the statements aloud (e.g. 8 is greater than 5; 1 is less than 3).

a. $9 > 8$ **b.** $5 < 7$ **c.** $1 < 3$ **d.** $4 > 2$

4. Draw a line to match **equal** (the same) amounts of circles.

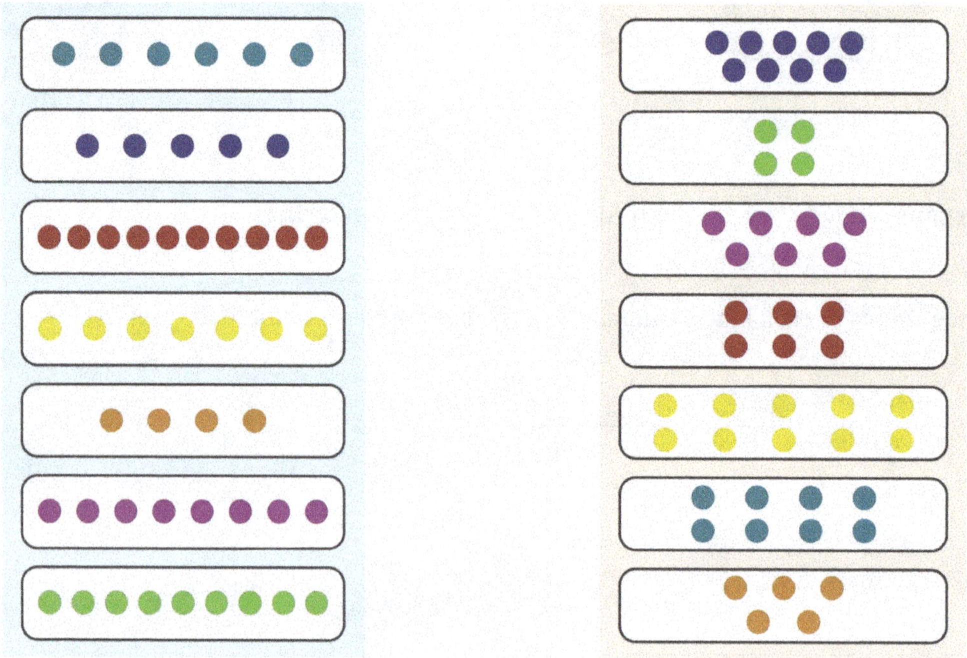

5. Note: the sign = means **equals**. Circle the number that is less. If the numbers are equal, circle both. Practice saying the statements aloud.

a. $6 < 9$ **b.** $4 = 4$ **c.** $5 > 3$ **d.** $8 > 7$

6. Circle the greater number. If the numbers are equal, circle both.

a. 5 2	**b.** 9 9	**c.** 1 1	**d.** 4 8
e. 6 7	**f.** 10 9	**g.** 2 3	**h.** 10 10

Sort Some More

1. **Activity.** (Optional.) Do the "Sort, and List by Count" activity (see chapter introduction). This is the same activity as in the previous lesson on sorting, so use your judgment to decide whether your child or students need it.

2. **a.** Count each **kind** of flower. (There are varying sizes of each kind of flower. You can identify them by color.)

_____ blue flowers _____ red flowers _____ yellow flowers _____ pink flowers

b. Write the count of each kind of flower **in order**, from the least count to the greatest.

_____ ____________________ flowers

_____ ____________________ flowers

_____ ____________________ flowers

_____ ____________________ flowers

Puzzle Corner	Write the numbers in order, from smallest to greatest.
a. 2 8 10 7	**b.** 5 3 6 8

Mixed Review Chapter 2

1. Write the missing numbers in the number path.

		4	5		7			

2. Cross out one of the objects so that there is **one less**. Write how many are left.

a. ________ **b.** ________ **c.** ________ **d.** ________

3. Draw more dots of a different color to make the given total. Write the counts of dots of each color.

a. 9	**b.** 5	**c.** 7
	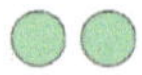	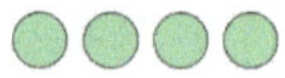
____ and ____	____ and ____	____ and ____

4. Color the picture following the color key.

1 = gray	3 = brown	5 = green	7 = dark blue	9 = light blue
2 = pink	4 = orange	6 = red	8 = yellow	10 = purple

Chapter 2 Review

Play "Put More in the Cup." (See the chapter introduction).

1. Match **equal** amounts of triangles.

2. Circle the number that is greater (more).

a. 9 8	**b.** 1 3	**c.** 7 10	**d.** 5 2

3. Draw a line to separate the dots in two groups, in different ways. Write how many are in each group.

a. 7 ○ ○ ○ ○ ○ ○ ○ _____ and _____	**b.** 7 ○ ○ ○ ○ ○ ○ ○ _____ and _____
c. 7 ○ ○ ○ ○ ○ ○ ○ _____ and _____	**d.** 7 ○ ○ ○ ○ ○ ○ ○ _____ and _____

Chapter 3: Addition & Measurement
Introduction

The third chapter of *Math Mammoth Kindergarten Workbook A* focuses first of all, on the concept of addition, and then secondarily on some foundations for measurement.

We start with the idea that in addition, we put two groups of objects together, and find the total. Then the lessons introduce the symbols + and =. In these first lessons, children are expected to find the sums by counting both groups of objects. But right after that, the lessons *Adding One More* and *Counting On* zero in on the important principle of counting on: there is no need to start counting from one when adding two numbers. For example, to add 4 + 3 by counting, you can count on from 4, three steps.

The number zero is dealt with in a separate lesson, to make sure children can grasp this concept.

Learning to solve word problems is important. We start with the simplest types of problems right here in kindergarten, and children will solve word problems with increasing difficulty in subsequent grade levels. At this level, you as the parent/teacher will probably need to read the problems aloud.

For that purpose, the chapter also includes some photographs of animals, with the idea that the child tells a story about the picture that uses math. We're thinking specifically of addition here, but at the same time, we're not looking for a specific, pre-determined story. So, don't reject what the child comes up with. For example, if in the story something goes away, you can teach the child about subtraction right there.

You can find more photos for such "math talks" at https://www.mathmammoth.com/mathtalks

Understanding what is happening in the problem is crucial. Encourage the child to use objects or drawings to help them make sense of the situation. If the child has difficulty, check that they understand all the words in the problem. Sometimes not knowing just one word throws them off. You can also try rewording the problem.

While this chapter has many lessons about addition, children do not need to memorize all addition facts at this stage. The goal for kindergarten is only that children learn to add fluently within 0-5, and they can take the entire kindergarten year to accomplish that. Then in first grade, children work on fluently adding and subtracting within 0-10, and in second grade, within 0-20.

In the lessons *Adding to Make 6*, *Adding to Make 7*, and so on, the child finds different decompositions of a number. For example, they will find different ways to add two numbers so the sum is 6, and they record these decompositions using drawings or with an equation (e.g. 7 = 3 + 4).

Again, these decompositions don't need to be memorized at this stage, but this work lays a foundation for first and second grades, when children will work on memorizing the basic addition facts based on decompositions. Also, decomposing a number into two parts helps children learn whole-part relationships with numbers.

The worksheet maker at https://www.mathmammoth.com/private/Make_extra_worksheets_K.htm allows you to make extra worksheets for decompositions and for addition.

With measurement, the goal is to understand that objects have the attributes of length, height, and weight, and that the "amount" of length, height, or weight varies among objects. (Weight is covered in chapter 5.) Children compare the length, height, and weight of objects informally, without actually using a measuring unit. They also describe several measurable attributes of the same object. For example, they might describe a pencil as being both light and long.

I have also included an optional lesson where children measure the length of objects with the nonstandard unit of paperclips. This helps children learn the basic idea in measurement, which is that we repeat the basic unit until reaching the length of the object being measured.

Please also see the following pages for games that I recommend using while studying this chapter. Games are important at this level, as they help children to practice addition and also make math fun.

Good Mathematical Practices

- This chapter introduces word problems. Naturally, they are very easy, but even here in kindergarten, we can help children to become good problem-solvers. One of the major attributes of good problem-solver is that they don't give up easily. So, don't help the child too early. Let them struggle a little bit.

  Here are some ideas to (eventually) help a child that is stuck. First make sure they understand what the problem says. You can re-read the problem to them and ask them then to recite the problem back to you. You can ask simple questions related to the facts given in the problem, such as, "How many chickens are there?" Then, you can ask the child to model the story with objects or to draw.

Pacing Suggestion for Chapter 3

Each lesson is intended to take one day. Feel free to take a day at times to just do activities and play games, without any written work. We want children to enjoy doing math, and not feel pressured.

The Lessons in Chapter 3	page	span	suggested pacing	your pacing
Adding as "Putting Together"	73	*1 page*	1 day	
Adding One More	74	*1 page*	1 day	
Addition Concept	75	*1 page*	1 day	
Counting On	76	*1 page*	1 day	
Adding Two More	77	*1 page*	1 day	
The Number Zero	78	*1 page*	1 day	
Sums Within 0-5, Part 1	79	*1 page*	1 day	
Addition on a Number Path	80	*1 page*	1 day	
More Addition on a Number Path	81	*1 page*	1 day	
Addition Word Problems	82	*1 page*	1 day	
Sums Within 0-5, Part 2	83	*1 page*	1 day	
Adding to Make 6	84	*1 page*	1 day	
In Which Order Do You Add?	85	*1 page*	1 day	
Shorter and Longer	86	*1 page*	1 day	
Adding to Make 7	87	*1 page*	1 day	
Shorter and Taller	88	*1 page*	1 day	
Adding to Make 8	89	*1 page*	1 day	
Comparing Lengths and Heights 1	90	*1 page*	1 day	
Comparing Lengths and Heights 2	91	*1 page*	1 day	
Adding to Make 9	92	*1 page*	1 day	
Measuring Length (optional)	93	*1 page*	1 day	
Adding to Make 10, Part 1	94	*1 page*	1 day	
Adding to Make 10, Part 2	95	*1 page*	1 day	
Mixed Addition Practice	96	*1 page*	1 day	
Mixed Review Chapter 3	97	*1 page*	1 day	
Chapter 3 Review	98	*1 page*	1 day	
TOTALS		*25 pages*	25 days	
with optional content		*(26 pages)*	(26 days)	

Games and Activities

Any ***board game*** where you move the piece by rolling two dice works to practice addition.

Bean Bag Toss *(Counting on from a number that is not 1)*

You need: Bean bags, 10 sheets of paper, a marker pen, and a dice. Write the numbers from 1 to 10 on the sheets of paper (one number per sheet), to be the targets. If the child knows how to write numbers, they could write them.

Activity: Place the sheets of paper on the floor or on stairs, etc., in scattered order. The child will roll the dice to get the starting number for counting. They will toss the first bean bag on the number they got from the dice, let's say 5, saying aloud, "Five!" Then they toss two to four more bean bags on the following numbers (on 6, 7, 8, 9), saying each number aloud as they toss the bag.

Addition Challenge *(Addition, comparing numbers)*

You need: Number cards with numbers from 1 to 5. (Regular playing cards will work.)

Game play: In each round, each player is dealt two cards face up, and calculates the sum. The player with the highest sum gets all the cards from the other players. After enough rounds have been played to use all of the cards, the player with the most cards wins. If two or more players have the same sum, those players get an additional two cards and use those to resolve the tie.

Note: when using regular playing cards, children can count the symbols on them to find the answer. (Encourage them to count *on* from the bigger number, and not start at 1.)

Variation: Instead of cards with numbers 1-5, use cards with 1-6 or some other range.

Animal Rescue with Addition *(Addition)*

You need: Two six-sided dice. Or, if you have them, two four-sided dice will make for an easier game. Number cards could also be used. About two dozen stuffed animals, plastic animals, lego figurines, bear counters, or similar things. In a pinch, use beans or marbles, and ask the child(ren) to imagine they are some type of animals. An environment for the animals where they are "in danger." For example, this could be a blanket that represents seawater, or it could be something representing a barn on fire, etc.

Game play: On your turn, roll the two dice, and "rescue" as many animals as is the total of the dice. Those animals become yours. In the end of the game, the player with the most rescued animals wins.

Domino Crabs *(Addition)*

You need: Seven markers for each of the two players. A set of dominoes from which you remove the ones where the sum of the dots is 7 or more. (Keep the dominoes where the sum is from 0 to 6.)
Download the game board from here:

https://www.mathmammoth.com/download/Domino-Crabs.pdf

Game play: Place the dominoes face down as if in a pond in between the players. Place the game board in between the players also. On your turn, draw one domino from the pond and add the dots. Then place a marker at your number on the board that is the sum. If you cannot do so, you will skip your turn. The first player to place all their markers on the crabs is the winner.

Addition Go Fish (*Addition*)

You need: The deck of cards for this game, which you can download from here:
https://www.mathmammoth.com/download/Addition-Go-Fish.pdf

Game play: Deal each player four cards. Spread the rest of the cards face down on the table as if a "pond." On your turn, first try to find matches (such as 5 + ● and 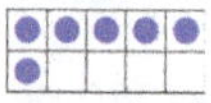) within your own cards. Remove any matches, placing them in your personal pile. Then, you may ask the player on your left for one specific number, such as, "Do you have an eight?" If the player has it (or one of the representations of 8), they have to give it. Otherwise, they say, "Go fish," and you will then take one card from the pond.

If you now get a match, you will again place those cards in your personal pile. Your turn will continue, and you can again ask the player on your left for one specific number. Once you cannot make a match with the cards, it is the next player's turn.

Continue playing until the pond is used up and no one can make any more matches. The winner is the player with most cards in their personal pile.

Zebra Dash Addition (*Addition within 0-5*)

You need: A six-sided die and a marker for each player. Download the game board from here:

https://www.mathmammoth.com/download/Zebra-dash-addition.pdf

Game play: First, each player rolls the die, and the player with the highest number starts the game. On your turn, roll the die, and move your marker that many spaces. If that square has instructions on what to do, follow those. If it has an addition problem, answer that. If your answer is wrong, move your marker backward one space.

Every time you come to a square with an addition problem, follow the same process: say aloud the answer and if your answer is wrong, move your marker backward one space. And if it has other instructions, follow those.

For example, let's say you answer 1 + 3 wrong, and go backward one space. That square states you move ahead one step, so you do so, and then you answer 1 + 3 again.

The person who reaches the watering hole first is the winner

Some Went Hiding (*Decomposing numbers*)

You need: Up to 10 small objects, such as marbles, pennies, beans, or rocks.

Game play: Choose a target number, such as 7. The first player shows seven objects to the next player, and then quickly hides *some* of them behind their back (without showing how many). The player then shows the rest of the items to the next player, who has to say how many "went hiding." If that player gives the right answer, it is then their turn to hide some and ask the next player to answer. If they give a wrong answer, they forfeit their turn. This game appeals best to young children.

Variation: Instead of getting a turn to hide objects, the player who answers correctly may gain points or other rewards for the right answer.

10 Out (or *5 Out* or *6 Out, etc.)* *(Number bonds)*

Note: This game is optional. It is challenging for kindergarten level. You can make it easier by providing manipulatives/objects or instructing the players to use their fingers.

You need: Lots of number cards from 1 to the target number.

Preparation: Choose a target number (sum) for the game, such as 10. Deal five cards to each player. Place the rest face down in a pile in the middle of the table.

Game play: On your turn, you may discard the card 10 (the target number) or any two cards in your hand that add up to it. You may also ask for one card from the player to your right (as in "Go Fish"), and if that player has it, they must give it to you. If you can discard all cards from your hand this way, you're the winner and the game ends. If not, take one card from the deck, and that ends your turn.

Variations:
* Deal more than five cards.
* Deal fewer cards if there are a lot of players or the players are very young.
* Allow players to discard *three* cards that add up to the target number.

Number Bonds in the Pond *(Number bonds)*

Note: This game is optional. It is challenging for kindergarten level, because the players don't have physical objects to determine the answer, but only see the abstract number shown on the card. You can make it easier by providing a manipulative or instructing the child to use their fingers.

You need: A standard deck (or several) of playing cards or number cards

Preparation: Choose a target sum for the game. If the target sum is 5, make a deck of cards consisting of numbers 1 through 4. If the target sum is 6, make a deck of numbers 1-5. And so on. (The deck always consists of numbers that are from 1 through X − 1 where X is the target sum.) Place a target number card face up between the players, and spread out the rest of the cards face down, like a pond, between the players.

Game play: On your turn, if you don't have any cards in your hand, take two cards from the pond. If you do, take one card from the pond. Now check if any two cards in your hand add up to the target number. If so, put those cards away in your personal pile. If not, it is the next player's turn. The game ends when there are no more cards in the pond. The winner is the person with the most cards in their personal pile.

Variation: Allow three cards/numbers to be added to reach the target number.

Notes: Depending on the number of players, you may need several decks of cards to make the pond. When first playing this game with your child/students, start with 5 as the target number, and advance to target numbers of 6, 7, and onward. Playing this game several times will help the child to memorize the number bonds (the sums) associated with a particular target number.

Math Mammoth Practice Zone

Addition Hidden Picture Game

Solve addition problems within 0-6 while uncovering a hidden picture in this fun game that children really like!
https://www.mathmammoth.com/practice/mystery-picture#min=0&max=6

"7 Up" card game

You will see seven cards dealt face up. Simply choose any two cards that make 10 (or your chosen sum) to discard.
https://www.mathmammoth.com/practice/seven-up

Further Resources on the Internet

These resources match the topics in this chapter, and offer online practice, online games (occasionally, printable games), and interactive illustrations of math concepts. We heartily recommend you take a look. Many people love using these resources to supplement the bookwork, to illustrate a concept better, and for some fun. Enjoy!

https://links.mathmammoth.com/Kch3

Scan me

Adding as "Putting Together"

Emily has four cookies and Megan has three.

They put the cookies together.
How many cookies are there in total?

Count them! There should be 7 cookies in total.

Together: 

1. **Activity: "Put together".** Work in pairs (parent/child, teacher/child, or two children). One person shows the other two distinct groups of objects (marbles, buttons, etc. — the sum should be 10 or less). The other counts the objects in each group. Then they put the objects *together* and find how many there are in total. Repeat, taking turns. Optionally, record at least one such addition on paper or whiteboard, using pictures.

2. Solve. You can use objects or draw pictures to illustrate the problems.

a. Katrina picked two flowers and Anita picked five.

The girls put their flowers together. How many flowers do they have in total? _______

b. Justin collected six seashells at the beach and Kim collected three.

How many seashells do they have in total? _______

c. Alex has four marbles and Josh has six marbles.
They put their marbles in the same bag.
How many marbles are in the bag now?

d. Kyle has three pencils and Emma also has three pencils.
How many pencils do they have together?

3. Make up a story from this picture that uses math and numbers.

Adding One More

1. Count the circles. Then draw **one more** circle. Tell or write the *new* total on the line.
(See if the child can say the new total without counting all the circles again.)

a.	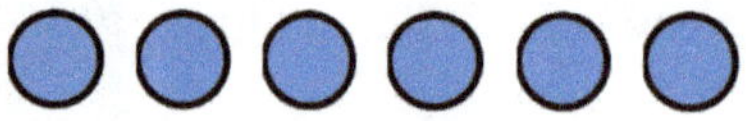	______
b.	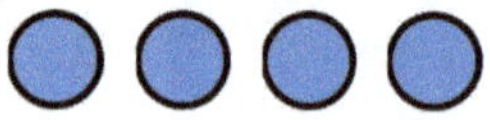	______
c.		______

2. Solve. Draw the answer, or show the answer with objects. Note: The child can simply draw circles or sticks.

a. Clayton had two oranges. Then his mom gave him one more.
How many oranges does Clayton have now?

b. Layla picked three red flowers and two yellow ones.
How many flowers did Layla pick in total?

c. Eight kittens were in a box. Then
another kitten jumped into the box.
How many kittens are in the box now?

3. Tell an addition that matches the picture.

4. If you draw one more pear, then
how many pears will there be?

Addition Concept

We use the plus sign (+) to show that two numbers are **added**.

We **add** (or put) the two groups of cookies together, and we get a total of 8 cookies. Since 2 and 6 together **equals** 8, or **is the same** as 8, we use the equals sign =, and write $2 + 6 = 8$.

$2 + 6 = 8$

Two plus six equals eight.

1. Add. Write the total in the box.

a. $5 + 2 = \square$

b. $4 + 1 = \square$

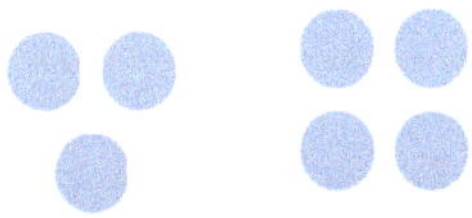

c. $3 + 4 = \square$

d. $6 + 2 = \square$

2. Write how many there are of each kind of fruit. Then write the total in the box.

a. ______ + ______ = $\square$

b. ______ + ______ = $\square$

3. Write the numbers. Add. Write the total in the box.

a. ______ + ______ = $\square$

b. ______ + ______ = $\square$

c. ______ + ______ = $\square$

d. ______ + ______ = $\square$

Counting On

To find the total number of objects in two groups, you can start with the number of objects in the first group, and **count on from that number** to find the total of all objects.

Count (say): 4, 5, 6, 7.
(Do not count from 1.)

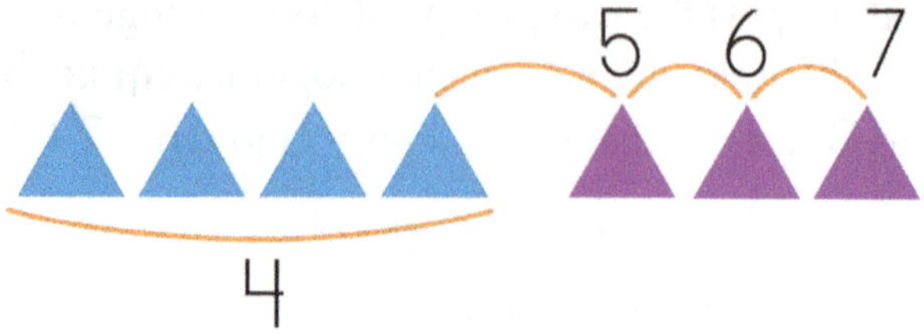

1. Starting with the total of the first group of objects, count on to find the total number of objects. (Count aloud with your teacher.)

5

3

2. Counting on from the number, count the dots. (Count aloud with your teacher.)

4 ●●●	2 ●●●●●	6 ●●●

3. **Activity.** Work in pairs (child with a teacher/parent/another child). Give the child a number (e.g. 3), and ask the child to count three steps forward *from* that number (not from 1). Take turns. **Note:** If the child or students do not know the counting sequence beyond 10, don't use numbers greater than 7 as starting points.

4. Count on two steps from the given number.

a. 5 ____ ____ **b.** 2 ____ ____ **c.** 8 ____ ____

5. Count on three steps from the given number.

a. 7 ____ ____ ____ **b.** 1 ____ ____ ____ **c.** 3 ____ ____ ____

6. Count the kiwis. Then draw **one more** kiwi. Tell or write the *new* total on the line.

a.  ______

b.  ______

7. Make your own color pattern, and continue it.

Adding Two More

1. Count the circles. Then draw **two more** circles. Tell or write the *new* total on the line.
(**Hint:** prompt the child to **count on** from the original number of circles to find the new total.)

a.	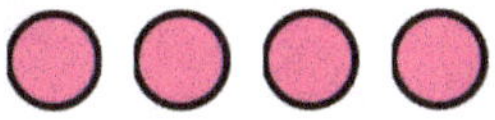	________
b.	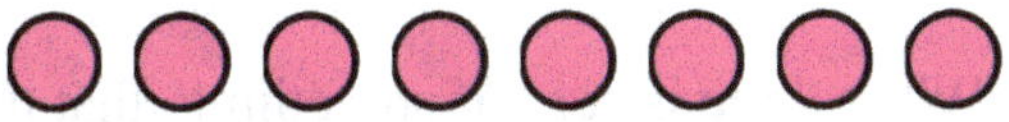	________

2. Tell a story about this picture that uses addition.

3. Solve. Draw something (dots, sticks, etc.) to show the answer.

a. Clayton had 7 toy cars. Then he bought two more. How many toy cars does he have now?
b. Five kittens were drinking milk. Then a puppy came to drink it, too. Now how many animals are drinking milk?
c. Four children were playing a game. Then Abby and Matt joined them. How many people are playing the game now?

4. Add. Write the total in the box.

 **a.** 4 + 3 = ☐	 **b.** 5 + 5 = ☐

The Number Zero

This is the number zero: 0. It comes before the number one.

0 1 2 3 4 5

Notice that the box above zero is empty. When there is zero of something, that means there is nothing.

1. Count out loud with your teacher, using your fingers to illustrate each number. (Practice counting at least twice.)

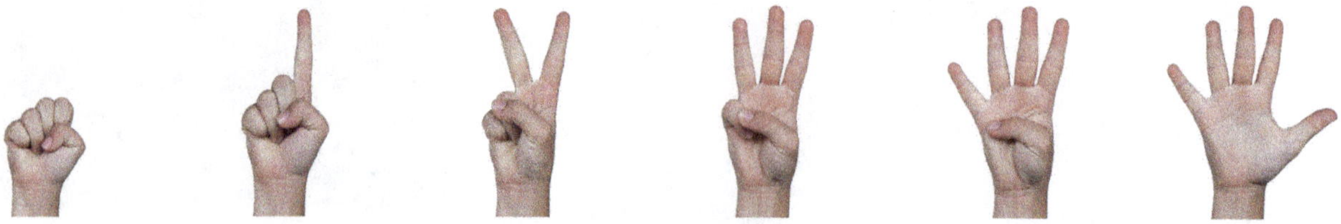

2. Circle all of the zeros in this box of numbers.

0	3	6	4	0	1	0

3. **a.** Circle the rabbit that has **zero** carrots.

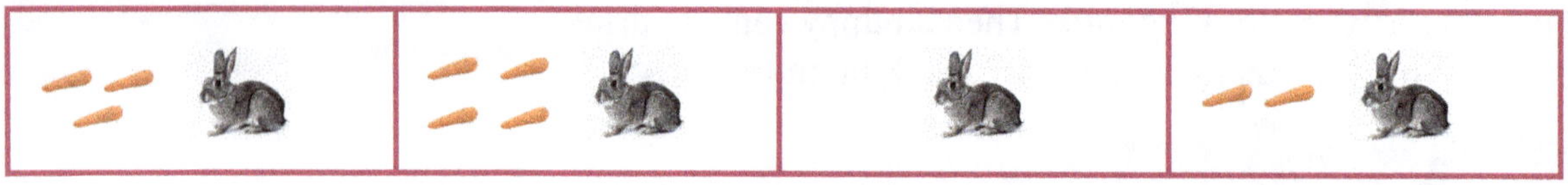

b. Circle the cat that has zero kittens.

4. Write how many flowers are in each box.

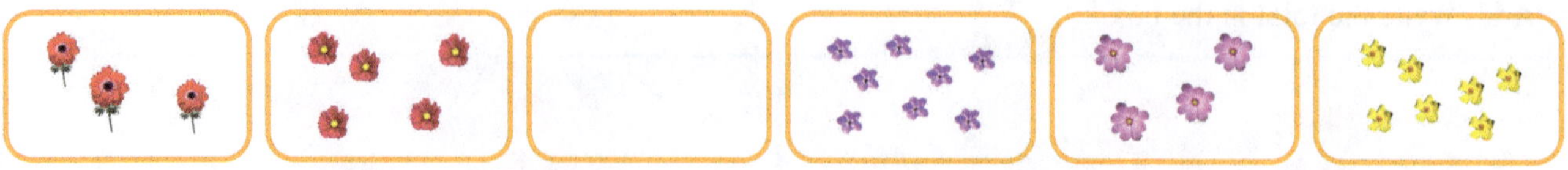

_____ _____ _____ _____ _____ _____

Sums Within 0-5, Part 1

1. **Game.** Play "Some Went Hiding" with 5 objects. (See the chapter introduction.)

2. Write how many there are of each kind of object. Add and write the total in the box.

a. 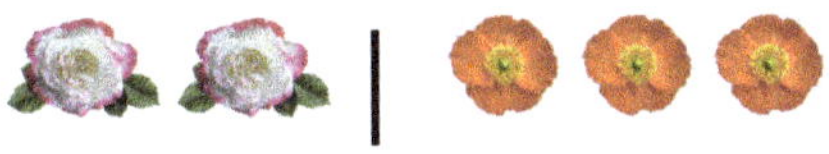

2 + 3 = ☐

b. 

____ + ____ = ☐

c. 

____ + ____ = ☐

d. 

____ + ____ = ☐

e. 

____ + ____ = ☐

f. 

____ + ____ = ☐

When we add zero to a number, we are adding **nothing**. The number doesn't change.

3 + 0 = 3

3. Practice adding with zero.

a. 

1 + 0 = ____

b. 

____ + ____ = ____

c. 

____ + ____ = ____

d. 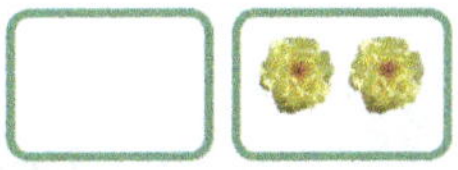

____ + ____ = ____

Addition on a Number Path

$4 + 2 = ?$ We can use a number path:

The last number tells you the total: $4 + 2 = 6$

1. Use the number path below to help you add.

1	2	3	4	5	6	7	8	9	10

a. $5 + 2 =$ ______	**b.** $3 + 2 =$ ______	**c.** $4 + 3 =$ ______
d. $1 + 3 =$ ______	**e.** $7 + 3 =$ ______	**f.** $3 + 5 =$ ______

2. Write the addition sentence.

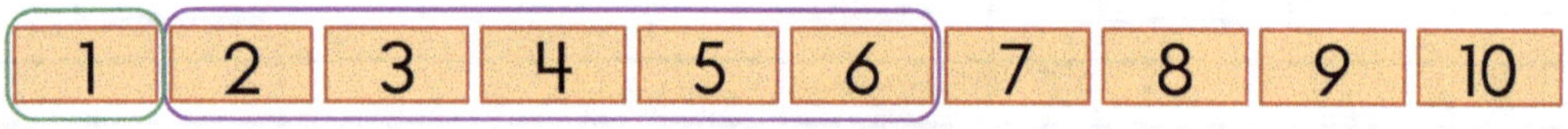

a. ________ + __________ = _________

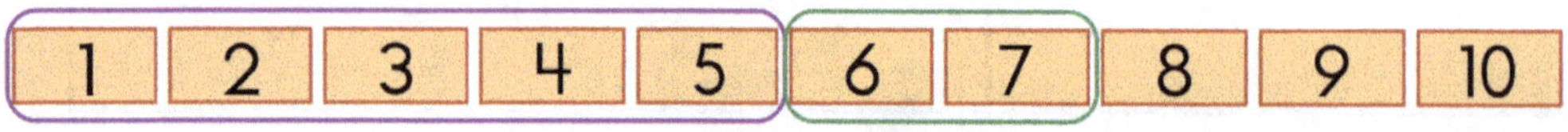

b. ________ + __________ = _________

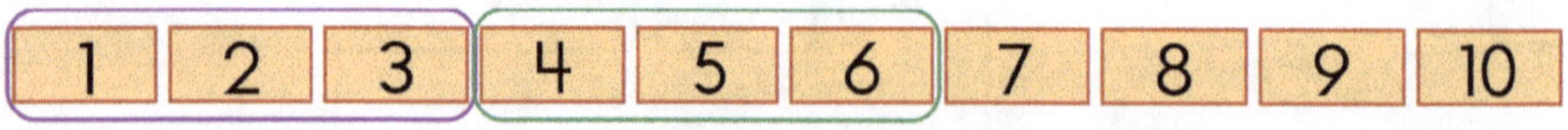

c. ________ + __________ = _________

3. **Activity.** Work in pairs (child with a teacher/parent/another child). Count on three steps (orally) from the number that the other person gives you. Or, use the numbers below as starting points. Take turns.

a. 4 **b.** 1 **c.** 7 **d.** 5 **e.** 3

More Addition on a Number Path

1. What happens if you add zero to a number? Or if the first number is zero? How would you illustrate it on a number path? Discuss with your teacher.

1	2	3	4	5	6	7	8	9	10

a. 5 + 0 = _____	b. 8 + 0 = _____	c. 0 + 3 = _____

2. **Activity.** Work in pairs (child with a teacher/parent/another child). Count backward three steps (orally) from the number that the other person gives you. Or, use the numbers below as starting points. Take turns.

a. 5 **b.** 8 **c.** 6 **d.** 10 **e.** 3

3. Write the addition sentence.

1	2	3	4	5	6	7	8	9	10

a. _______ + _________ = ________

1	2	3	4	5	6	7	8	9	10

b. _______ + _________ = ________

4. Add. You can use the number path to help.

1	2	3	4	5	6	7	8	9	10

a. 7 + 3 = _____	b. 3 + 4 = _____	c. 2 + 5 = _____
d. 3 + 4 = _____	e. 2 + 4 = _____	f. 5 + 4 = _____

5. Let's count backward! Complete.

a. 4 3 _____ _____	b. 7 _____ _____ 4 3

Addition Word Problems

1. Nicky had three stuffed animals. Her mom gave her some more.
   How many stuffed animals does she have now?
   (Count on from three to find the total.)

3

2. Draw pictures to illustrate the problems. Tell the answers to your teacher.

a. David has four strawberries and Candice has: 

They put their strawberries together. How many strawberries do they have in total?

b. Mia has seven crayons and Carol has three. They put their crayons together. How many crayons do they have in total?

c. Susan ate five crackers. Then she ate one more. How many crackers did she eat in total?

3. Continue the number patterns.

a. 5 6 7 5 6 7 ____ ____ ____ ____ ____ ____ ____

b. 1 4 1 4 ____ ____ ____ ____ ____ ____ ____ ____

Sums Within 0-5, Part 2

1. Write how many there are of each kind of object. Add and write the total on the line.

a. 

_____ + _____ = _____

b. 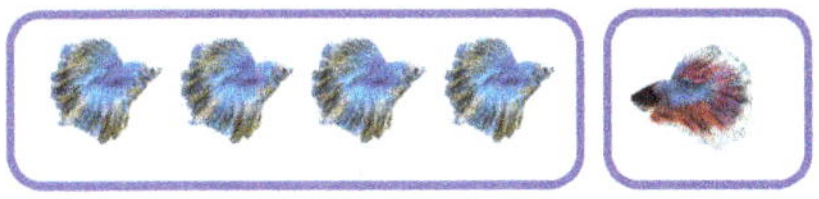

_____ + _____ = _____

c. 

_____ + _____ = _____

d. 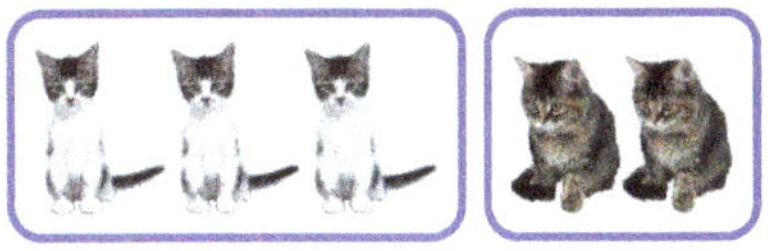

_____ + _____ = _____

2. Draw circles in the empty box to illustrate the number. Add.

a. ○ [ ]

$1 + 3 =$ _____

b. [ ] ○

$2 + 1 =$ _____

3. Kyle is having a pizza party for his birthday! Kyle's mom baked **four** pizzas.
Then she baked this many more:

How many pizzas did she bake?

4. Add. **Note:** The student can use objects, drawings, number path, an abacus, *etc.* to help solve the problems.

a. $3 + 1 =$ _____	b. $1 + 4 =$ _____	c. $0 + 5 =$ _____
$2 + 0 =$ _____	$2 + 3 =$ _____	$1 + 2 =$ _____

Adding to Make 6

1. **Activity**. Find all the ways to decompose 6. Use small objects or Cuisenaire rods. Record all those ways by drawing and/or by writing an addition equation in the table below or on your own recording sheet.

a.  6 = _____ + _____	**b.**  6 = _____ + _____
c.  6 = _____ + _____	**d.**  6 = _____ + _____
e.  6 = _____ + _____	**f.**  6 = _____ + _____

2. (optional) Referring to your work in question 1, think: How could you organize the decompositions in some specific order? (You can rewrite the decompositions on a separate paper, as a list.)

3. **Game.** Play "Some Went Hiding" with 6 objects.

4. Draw more dots to make a total of **six** dots. On the line, write how many dots you drew.

a. 2 + _____ = 6	**b.** 5 + _____ = 6	**c.** 3 + _____ = 6

5. **Count on** from the bigger number to add. Don't count from 1.

a. 3 + 2 = ______	**b.** 5 + 3 = ______	**c.** 8 + 2 = ______
7 + 1 = ______	5 + 1 = ______	6 + 1 = ______
6 + 2 = ______	4 + 2 = ______	6 + 3 = ______

In Which Order Do You Add?

1. Compare the two additions. What is the same? What is different?

a.  $1 + 4 =$ ____ OR  $4 + 1 =$ ____

b.  $2 + 3 =$ ____ OR  $3 + 2 =$ ____

c.  $5 + 2 =$ ____ OR 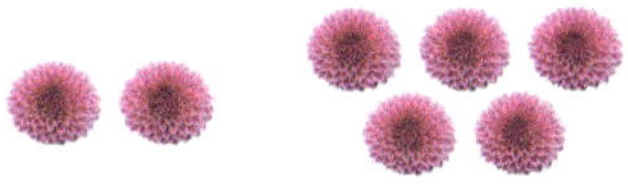 $2 + 5 =$ ____

It does not matter in which order you add two numbers. You can add in whichever order is **easier** for you.

Here, you can **count on** from the bigger number, the seven.

$1 + 7 =$ ____

or

$7 + 1 =$ ____

2. Add. Count on from the bigger number; don't count from 1.

a. $8 + 2 =$ ____	**b.** $6 + 1 =$ ____	**c.** $5 + 2 =$ ____

3. Add by counting on. Which order is easier? (Remember to not count from 1.)

a. $1 + 3 =$ ____ $3 + 1 =$ ____	**b.** $6 + 2 =$ ____ $2 + 6 =$ ____	**c.** $2 + 7 =$ ____ $7 + 2 =$ ____
d. $8 + 1 =$ ____ $1 + 8 =$ ____	**e.** $2 + 4 =$ ____ $4 + 2 =$ ____	**f.** $6 + 3 =$ ____ $3 + 6 =$ ____

Shorter and Longer

Look at these crayons. Which one is longer? Which one is shorter?

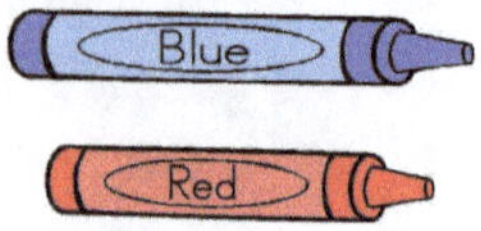

The blue crayon is **longer** (than the red). The red crayon is **shorter**.

1. **Activity.** Gather together items of varying lengths, such as pencils, toys, pieces of string, Cuisenaire rods, sticks, *etc.* (Note: Repeat each of the activities below several times with different items.)

   a. Show the child a short object and tell them it is short. Do similarly for one long object. Then ask the child to describe the length of some other object. For example: "Based on what we've talked about, how would you describe *this*?" (The answer could be: "It is long.")

   b. Set two objects beside each other and ask the child, "Which one is longer?", or "Which one is shorter?"

   c. Tell the child a comparison and ask them if it is true. For example, "This toy is longer than the spoon." Is that right?

   d. Choose an object. Ask the student to find an object that is shorter (or longer) than the chosen object.

   e. Set three objects beside each other and ask the child to find the longest or the shortest among them.

2. Circle the object that is longer.

a. 

b. 

c. 

3. Circle both the longest and the shortest object.

a. 

b. 

c. 

4. Add in the easiest order.

a. $1 + 7 =$ ______	**b.** $9 + 1 =$ ______	**c.** $2 + 6 =$ ______

Adding to Make 7

1. **Activity**. Find all the ways to decompose 7. Use small objects or Cuisenaire rods. Record all those ways by drawing and/or by writing an addition equation in the table below or on your own recording sheet.

 7 = ____ + ____	7 = ____ + ____
7 = ____ + ____	7 = ____ + ____
7 = ____ + ____	7 = ____ + ____

2. (optional) Referring to your work in question 1, think: How could you organize the decompositions in some specific order? (You can rewrite the decompositions on a separate paper, as a list.)

3. Solve the problems. You can draw pictures to help. Tell the answers to your teacher.

a. Amy has two limes and Liam has: . They put their limes together. How many limes do they have in total?	**b.** Ron ate: . Then he ate one more cookie. How many cookies did he eat in total?

4. Circle the number that is **greater**. Practice saying the comparisons aloud.

$0 < 2$ $3 < 5$ $9 > 8$ $4 > 2$ $6 < 9$

5. Add. (You can count on from the bigger number. Use drawings or objects only if necessary.)

a. 2 + 0 = ____	**b.** 3 + 2 = ____	**c.** 1 + 1 = ____
4 + 1 = ____	0 + 4 = ____	1 + 2 = ____

Shorter and Taller

Look at these two trees.
Which one is taller?
Which one is shorter?

1. **Activity:** Gather together objects of varying heights, such as toys, cereal boxes, canned foods, *etc.*

   a. Show the child a tall object and tell them it is tall. Do similarly for one short object. Then ask the child to describe the height of some other object. For example: "Based on what we've talked about, how would you describe *this*?" (The answer could be: "It is short.")

   b. Set two objects side-by-side. Ask the child, "Which one is taller?", or "Which one is shorter?"

   c. Tell the child a comparison and ask them if it is true. For example, "The bear is taller than the duck." Is that right?

   d. Ask the child to compare the heights of two objects. E.g. "Here we have a mug and a toy. How would you compare their heights?" (The answer could be for example: "The mug is taller.")

   e. Place three objects standing beside each other and ask, "Which one is the tallest?" (or shortest).

2. Circle the object that is shorter.

3. Draw pictures to illustrate the situations.

a. The girl is taller than the boy.	**b.** The blue box is shorter than the red box.

Puzzle Corner

How many more ducklings do you need, in order to have 10 total?

Adding to Make 8

1. **Activity**. Find all the ways to decompose 8. Use small objects or Cuisenaire rods. Record all those ways by drawing and/or by writing an addition equation in the table below or on your own recording sheet.

 8 = _____ + _____	 8 = _____ + _____
 8 = _____ + _____	 8 = _____ + _____
 8 = _____ + _____	 8 = _____ + _____

2. (optional) Referring to your work in question 1, think: How could you organize the decompositions in some specific order? (You can rewrite the decompositions on a separate paper, as a list.)

3. **Game** (optional). Play "Some Went Hiding" (see chapter introduction). If using 7 or 8 objects is challenging, use a number that the child or students are comfortable with, such as 4 and 5.

4. You already have five. Then you add to it. Use the number path to help you add.

1	2	3	4	5	6	7	8	9	10

a. 5 + 4 = _____ **b.** 5 + 5 = _____ **c.** 5 + 3 = _____

5. Add. Think which order is easier. (Hint: Count on from the bigger number.)

a. 1 + 9 = _____	**b.** 1 + 6 = _____	**c.** 5 + 2 = _____
d. 2 + 3 = _____	**e.** 0 + 7 = _____	**f.** 2 + 8 = _____
g. 1 + 7 = _____	**h.** 6 + 2 = _____	**i.** 1 + 4 = _____

Comparing Lengths and Heights 1

1. **Activity.** Gather a variety of items with varying lengths and heights, for the purpose of comparing their lengths and heights. Show the child two items and describe one as longer than the other. Also describe the other as shorter than the first one. (E.g. say, “The ruler is longer than the eraser. Can you see that? And the eraser is shorter than the ruler.”) Then ask the child to compare the lengths or heights of other pairs of items. The goal is for the child to use the expressions shorter than/longer than/taller than/shorter than in sentences.

   At your judgment, you can also expand this to include three items, and the expressions “longest of them all”, “shortest of them all”, etc.

2. Fill in (tell your teacher):

a. The _______ is shorter than the ________.

b. The _______ is longer than the ________.

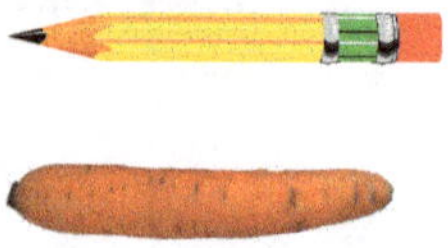

3. Which is true?

a. The shoe is shorter than the scissors.

or The shoe is longer than the scissors.

b. The toy car is shorter than the pen.

or The toy car is longer than the pen.

4. Compare the length of these things, and describe the difference. In other words, describe one item as <u>shorter</u> **or** <u>longer</u>. For example, “The cart is longer than the bike” or “The spoon is shorter than the fork.”

a.

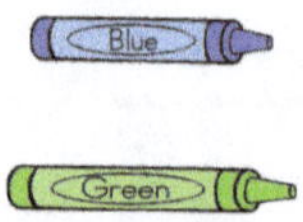

b.

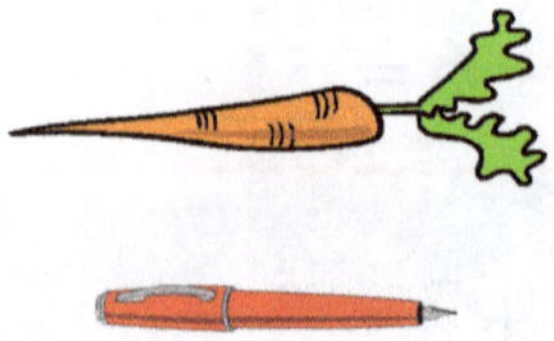

c.

5. Play an addition game (see the chapter introduction).

Comparing Lengths and Heights 2

1. Fill in (tell your teacher):

a. The ________ is shorter than the ________.

b. The ________ is longer than the ________.

2. Compare the height of these things, and describe the difference. In other words, describe one item as <u>taller</u> **or** <u>shorter</u>. For example, “The tree is taller than the car.”

a.

b.

Amy Sara

c.

1 2

3. Draw pictures to illustrate the situations.

a. The green string is shorter than the yellow string.

b. There are three pencils, and the orange pencil is longest of them all.

4. Add.

a.	**b.**	**c.**
$3 + 2 =$ ______	$1 + 3 =$ ______	$2 + 2 =$ ______
$4 + 0 =$ ______	$1 + 5 =$ ______	$3 + 3 =$ ______
$2 + 2 =$ ______	$0 + 3 =$ ______	$1 + 2 =$ ______

Adding to Make 9

1. **Activity**. Find all the ways to decompose 9. Use small objects or Cuisenaire rods. Record all those ways by drawing and/or by writing an addition equation in the table below or on your own recording sheet.

a. 9 = _____ + _____	**b.** 9 = _____ + _____
c. 9 = _____ + _____	**d.** 9 = _____ + _____
e. 9 = _____ + _____	**f.** 9 = _____ + _____

2. (optional) Referring to your work in question 1, think: How could you organize the decompositions in some specific order? (You can rewrite the decompositions on a separate paper, as a list.)

3. Let's count backward! Complete the sequences.

a. 4 3 _____ _____ _____	**b.** 9 _____ _____ _____ 5 4

4. How many more circles do you need to make the target number?

a. 5	**b.** 6	**c.** 7
●●●	●●	●●●●
3 + _____	2 + _____	4 + _____

5. Add.

a. $0 + 3 =$ _____	**b.** $2 + 4 =$ _____	**c.** $7 + 0 =$ _____
$2 + 3 =$ _____	$7 + 2 =$ _____	$1 + 2 =$ _____

Measuring Length

(This lesson is optional.)

In this lesson, we measure how long some objects are by comparing them to a paperclip. We will **count how many paperclips long** the objects are. This means we are using a paperclip as a **measuring unit**.

It is important to make sure that the paperclips you are using are all the same size. Also, make sure they are placed end-to-end (touching each other), with no gaps or overlaps.

The pencil is six paperclips long.

1. How many paperclips long are the pencils below? Tell the answers to your teacher.

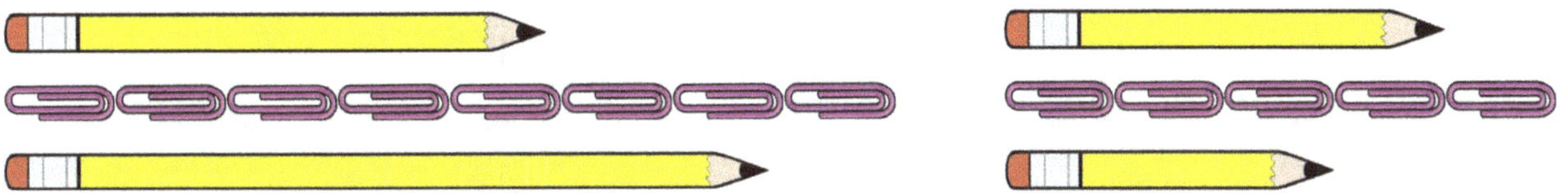

2. Measure how long some things are, using paperclips. **You need**: several paperclips that are the same size, small things to measure such as an eraser, a pencil, a fork, a ruler, crayons, toys, etc. Check beforehand that the objects you choose are spanned by a whole number of paperclips (with no gaps or overlaps). Optionally, you can write the names of the objects and their lengths below.

____________________________________ ______ paper clips

____________________________________ ______ paper clips

____________________________________ ______ paper clips

____________________________________ ______ paper clips

3. How many pencils long are the candle and the bread? Tell the answers to your teacher.

a.

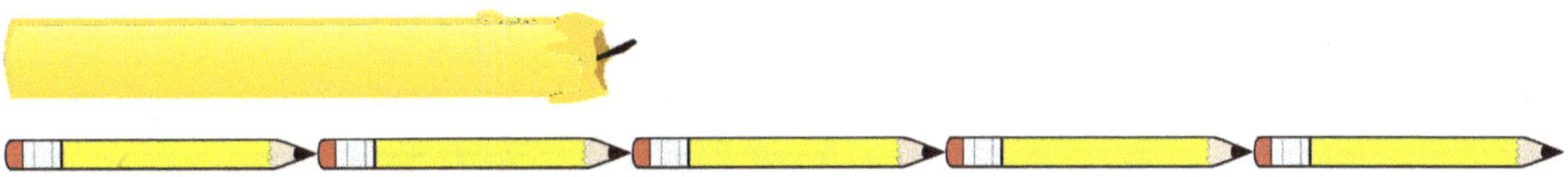

b.

Adding to Make 10, Part 1

1. **Activity**. Use Cuisenaire rods, ten objects, or fingers. Find all the different ways to make 10 with two numbers. Optionally, write the corresponding sums on blank paper. Can you find a pattern in the sums?

This is a **ten-frame**. In it, you can fit **ten** objects.

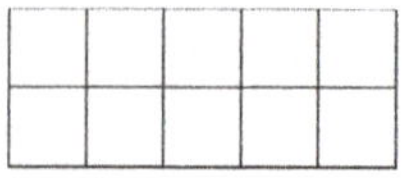

Here the ten-frame is filled with 5 red dots and 5 blue dots — a total of 10! 5 + 5 = 10.

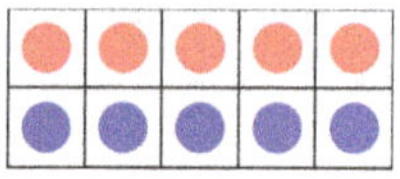

2. Add the red and the blue dots in the ten-frame. What addition can you write?

a. 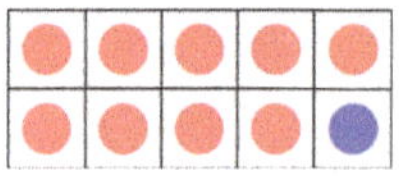

____ + ____ = ______

b. 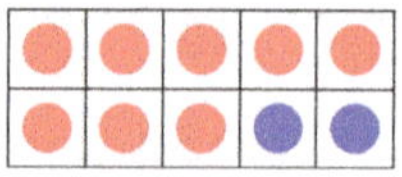

____ + ____ = ______

c. 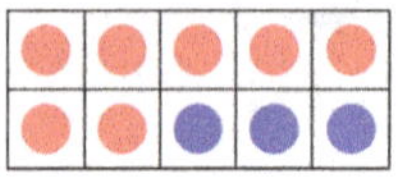

____ + ____ = ______

d. 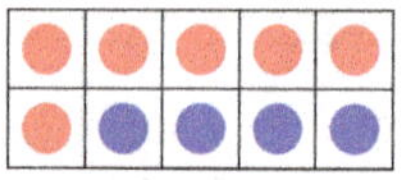

____ + ____ = ______

3. Find the missing number so that the total is 10. You can draw dots in the ten-frame to help you.

a. 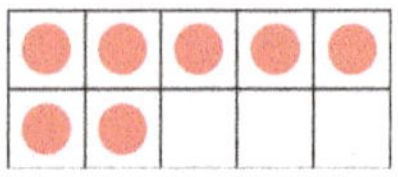

7 + _____ = 10

b. 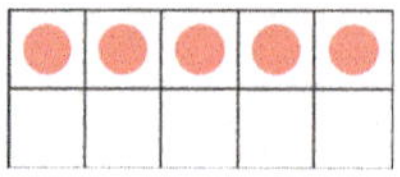

5 + _____ = 10

c. 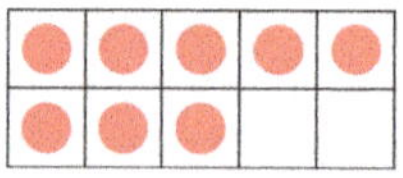

8 + _____ = 10

d. 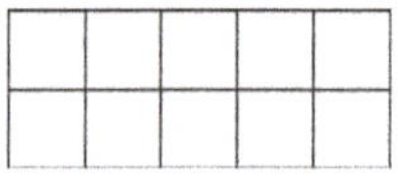

4 + _____ = 10

e. 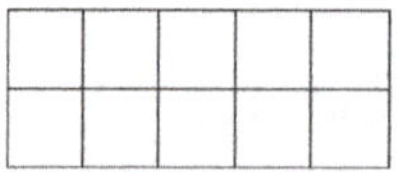

1 + _____ = 10

f. 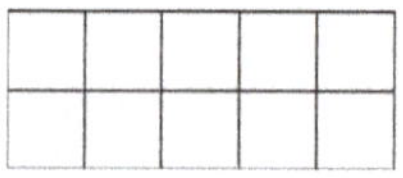

6 + _____ = 10

4. Write the number that is one less.

a. 9 ______ **b.** 6 ______ **c.** 7 ______ **d.** 1 ______

Adding to Make 10, Part 2

1. **Activity**. Print the page with ten-frames from the printables provided with the download, or from this link: https://www.mathmammoth.com/download/ten-frames.pdf
   Cut them out (leave out the duplicate of zero dots) and put them in a pile face down. Have the child pick a ten-frame from the pile, and tell how many more to add to it to make 10. Take turns.

   Alternatively, you can use Cuisenaire rods from 1 to 9 or 1 to 10. Hide them all in a hat or bag. Have the 10-rod visible for a reference and comparison.

2. Count down from 10 to zero.

3. Find the missing number so that the total is 10. You can draw dots in the ten-frame to help you.

a. 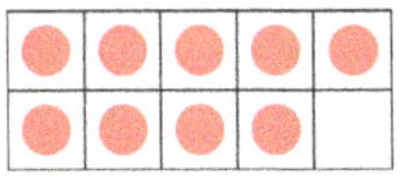

$9 + \_\_\_\_ = 10$

b. 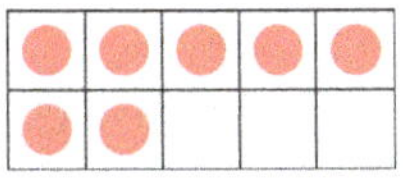

$7 + \_\_\_\_ = 10$

c. 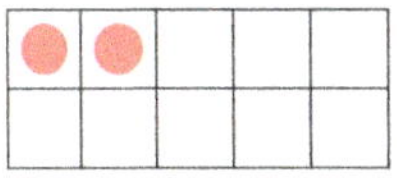

$2 + \_\_\_\_ = 10$

d. 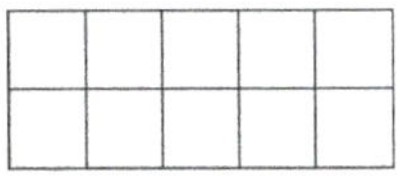

$8 + \_\_\_\_ = 10$

e. 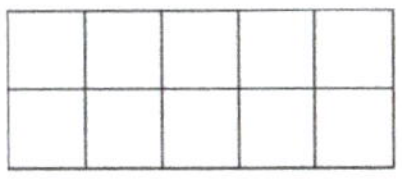

$5 + \_\_\_\_ = 10$

f. 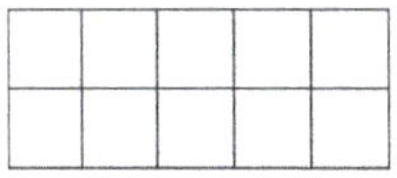

$4 + \_\_\_\_ = 10$

4. Circle the number that is **less**. If the two numbers are **equal**, circle both numbers.

$2 = 2$ $\qquad$ $9 > 5$ $\qquad$ $1 = 1$ $\qquad$ $3 < 7$

5. Find the missing numbers. You can draw, use the number path, or use other things to help you.

1	2	3	4	5	6	7	8	9	10

a. $3 + \_\_\_\_ = 7$ $\qquad$ **b.** $4 + \_\_\_\_ = 7$ $\qquad$ **c.** $6 + \_\_\_\_ = 10$

6. Circle the tallest object **and** the shortest object.

a. 

b. 

c. 

Mixed Addition Practice

1. **a.** Amy picked six flowers and Eve picked  . They put their flowers together. How many flowers do they have in total?

b. Tell a "math story" about this picture.

2. This is a new way to write addition! The answer goes *under* the line.

a. $\begin{array}{r} 2 \\ +\ 1 \\ \hline 3 \end{array}$	**b.** $\begin{array}{r} 4 \\ +\ 0 \\ \hline \end{array}$	**c.** $\begin{array}{r} 3 \\ +\ 2 \\ \hline \end{array}$	**d.** $\begin{array}{r} 4 \\ +\ 1 \\ \hline \end{array}$	**e.** $\begin{array}{r} 2 \\ +\ 2 \\ \hline \end{array}$
f. $\begin{array}{r} 1 \\ +\ 0 \\ \hline \end{array}$	**g.** $\begin{array}{r} 0 \\ +\ 3 \\ \hline \end{array}$	**h.** $\begin{array}{r} 1 \\ +\ 1 \\ \hline \end{array}$	**i.** $\begin{array}{r} 1 \\ +\ 3 \\ \hline \end{array}$	**j.** $\begin{array}{r} 2 \\ +\ 0 \\ \hline \end{array}$

3. Illustrate each addition in some way, and find the answer.

a. $7 + 3 =$ ______

b. $4 + 4 =$ ______

4. Draw more circles to make the number at the top. Write how many circles you drew on the line.

a. 8	**b.** 6	**c.** 10
●●●●●	●●	●●●
5 + ______	2 + ______	3 + ______

Mixed Review Chapter 3

1. Count each group of dots and write the total on the line. Then circle the number that is **less**. If the groups are equal, circle both numbers.

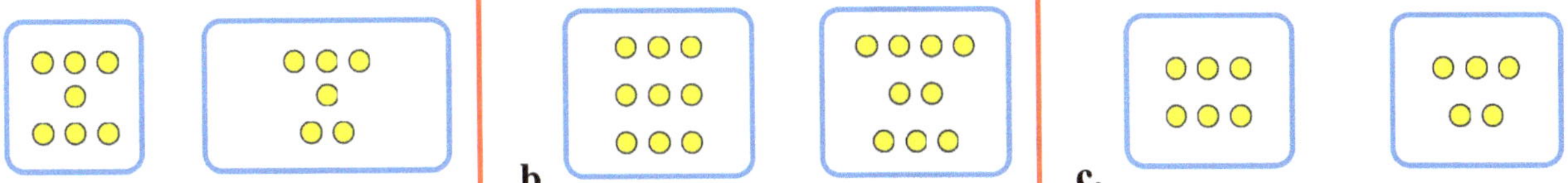

a. ______ ______ b. ______ ______ c. ______ ______

2. Write the missing numbers.

☐ 2 3 4 5 ☐ 7 ☐

3. In a., color the **longer** pencil. In b., color the **taller** animal. In c., color the **shorter** fish.

4. Draw another group of sticks so that there are **10 in total**. Write the numbers in each group.

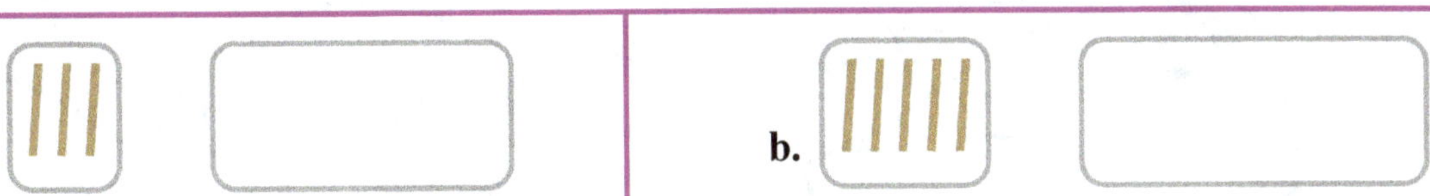

a. ______ and ______

b. ______ and ______

5. Add.

a.	b.	c.	d.	e.
5	3	4	1	2
+ 0	+ 1	+ 0	+ 1	+ 3

Chapter 3 Review

1. Compare the length or height of these things, describing one thing as taller, longer, or shorter.

a. 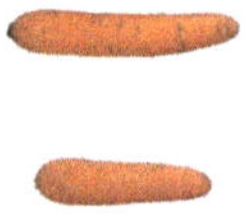

b. 

c. 

2. Draw pictures to illustrate the situations.

a. The green ruler is shorter than the yellow one.	**b.** The blue bird is taller than the brown bird.

3. Solve each addition, and color the sections with the colors that correspond to the answers.
Note: The student can use objects, drawings, an abacus, *etc.* to help solve the problems.

3 = yellow	5 = dark green	7 = blue	9 = white
4 = black	6 = light blue	8 = green	10 = any brown

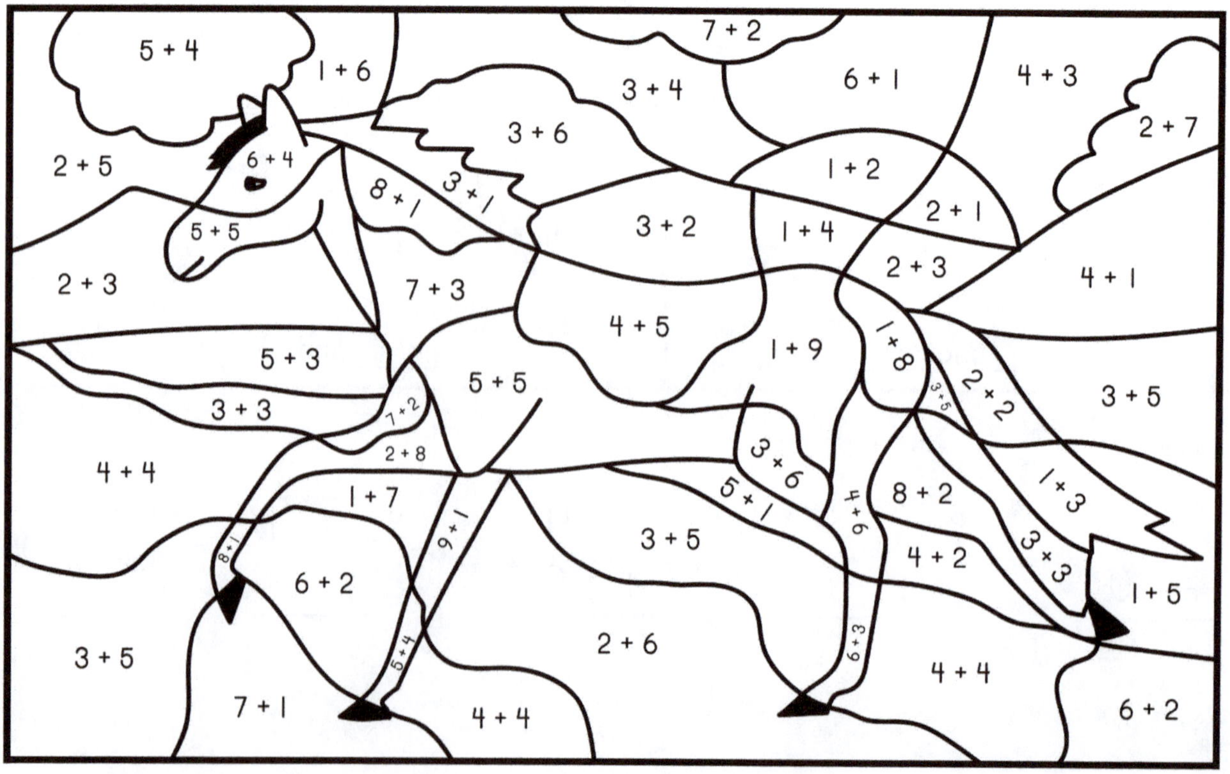

Chapter 4: Subtraction
Introduction

In chapter 4, we focus on subtraction. Our goals for the child(ren) are:

- understand subtraction as a process of "taking away";
- understand that when two parts make a total, you can subtract one part to find the other part (subtraction as "taking apart");
- tying in with the above, start learning the connection between addition and subtraction;
- be able to subtract within 0-10 with the help of pictures, objects, or a number path;
- be able to solve simple word problems involving subtraction;
- Fluently subtract within 0-5. This is the goal for the entire kindergarten year, not specifically for this chapter, but we want to work towards it here.

The first lesson presents subtraction as a process of "taking away." We work with that understanding for several lessons, and connect counting down with subtraction.

To help children count down to subtract, some lessons use a number path, which is similar to a number line, but is a simpler model for young children to understand. This is because it has actual items (rectangles) they can count, as opposed to a number line where we count the spaces (distances) *between* numbers.

The lessons on number bonds lead us to the thought of having two parts and a total (whole). When two or more parts make a whole, and you know the whole and the other part(s), you can subtract to find the unknown part. This is explained in detail in the lesson "Two Parts and a Total."

Yet another use for subtraction is in the concept of difference, which has to do with comparing two numbers, and finding out how much more (or less) one is than the other. We will not deal with that in kindergarten; it is studied in detail in first grade.

The lessons titled "Subtract from 6," "Subtract from 7," and so on provide more practice and help children understand the connection between addition and subtraction. Even though those lessons have to do with subtracting *from* 6 (or from 7, 8, 9, or 10), the subtractions in the exercises are often tied to the corresponding additions. Children will consolidate understanding this connection in first grade.

By the end of kindergarten, the goal is for the child to fluently add and subtract within 0-5. To that end, continue using the various games listed here (and others). For example, you can sometimes use the day's math lesson just to play games that develop fluency, or add a game to an existing lesson. You can also use the worksheet maker at https://www.mathmammoth.com/private/Make_extra_worksheets_K.htm to print addition and subtraction worksheets.

In first and second grades, students will greatly expand their addition and subtraction skills. They learn various strategies for addition and subtraction, solve word problems involving addition and subtraction, and solve missing number problems involving addition and subtraction. In first grade, students are expected to fluently add and subtract within 0-10, and in second grade, within 0-20.

Please see the following pages for games that I recommend using in this chapter.

Pacing Suggestion for Chapter 4

Each lesson is intended to take one day. Feel free to take a day at times to just do activities and play games, without any written work. We want children to enjoy doing math, and not feel pressured.

The Lessons in Chapter 4	page	span	suggested pacing	your pacing
Subtraction as "Taking Away"	104	*1 page*	1 day	
Subtracting One	105	*1 page*	1 day	
How Many Are Left?	106	*1 page*	1 day	
Subtracting Two	107	*1 page*	1 day	
Subtraction on a Number Path 1	108	*1 page*	1 day	
Subtraction on a Number Path 2	109	*1 page*	1 day	
Number Bonds 1	110	*1 page*	1 day	
Number Bonds 2	111	*1 page*	1 day	
Two Parts and a Total	112	*1 page*	1 day	
Two Parts and a Total — More Practice	113	*1 page*	1 day	
Subtraction within 0-5	114	*1 page*	1 day	
Subtraction within 0-5, Part 2	115	*1 page*	1 day	
Subtraction within 0-5, Part 3	116	*1 page*	1 day	
Subtract from 6	117	*1 page*	1 day	
Subtract from 7	118	*1 page*	1 day	
Word Problems	119	*1 page*	1 day	
Subtract from 8	120	*1 page*	1 day	
Subtract from 9	121	*1 page*	1 day	
Subtract from 10	122	*1 page*	1 day	
Mixed Review Chapter 4	123	*1 page*	1 day	
Chapter 4 Review	124	*1 page*	1 day	
TOTALS		*21 pages*	21 days	

Games and Activities

Minus One (Or Two) *(Subtracting One or Two)*

You need: A deck of number cards from 1-10.

Game Play: First, set a reasonable time limit for the players to say the answers, such as 2 or 3 seconds. The players sit in a circle, with the deck in the middle. On your turn, take one card from the deck, say the number on it, and the number that is one less, within the agreed time limit. For example, if you get the card with 5, you say, "Five, four." If you say this correctly, and within the time limit, you get to put the card in your personal pile. If not, the card goes back to the bottom of the deck.

Play until the deck is used up. The winner is the player with most cards in their personal pile; however, you don't have to necessarily find the winner.

Variation: Count back two steps instead of one.

Subtraction Icebergs *(Subtraction)*

This is a two-player game.

You need: Eight counters for each of the two players. A ten-sided die or one six-sided die and number cards from 1-10. Download the game board here:

https://www.mathmammoth.com/download/Subtraction-icebergs.pdf

Game Play: Each player places their eight counters on their eight icebergs. On your turn, to get two numbers to subtract, roll the ten-sided die twice, or roll the six-sided dice and draw one number card. Subtract the two numbers. Remove the counter from the iceberg that shows the answer to the subtraction. If you cannot go, you skip your turn. The player that removes all their counters from the icebergs first is the winner.

Subtraction Challenge *(Subtraction)*

You need: A deck of number cards from 1-10. You might also remove 10s, 9s, and 8s, using your judgment, in which case you may need two decks. Alternatively, a set of dominoes works well also. Limit it to dominoes with a maximum of 10 dots on one side.

Game Play: In each round, each player is dealt two cards (or 1 domino) face up, and has to calculate the difference (subtract). The player with the highest difference gets all the cards (dominoes) from the other players. After enough rounds have been played to use up the deck, the player with the most cards wins.

If two or more players have the same difference, then those players get an additional two cards (one domino) and use those to resolve the tie.

Let's Make a Difference! *(Subtraction)*

You need: A large deck of number cards with numbers 1-10. Optionally, some small objects, number path, or an abacus to help children to do the subtractions.

Preparation: Deal six cards to each player. If a player cannot hold the cards, they can place them in front of them on the table. Place the rest face down in a pile in the middle of the table.

Game play: On your turn, check if you can make a subtraction sentence using any three cards in your hand. If so, say that subtraction aloud, and put those three cards in your personal pile. If not, take *one* card from the middle pile, and check again. If you still cannot make a subtraction, your turn ends. At the end of your turn, draw enough cards from the middle pile so that you once again have at least six cards in your hand (if you already have six or more, don't draw more at this point). The turn then passes to the next player.

Play in this manner until there are no more cards in the middle pile. At that point, the player with most cards in their personal pile wins.

Note: Essentially, players can check if they have three cards that make an *addition* sentence. However, it is best if the child notices that on their own.

Variation: For an easier game, remove some of the higher numbers from the deck. You can also deal each player five cards, instead of six; however this will make it less likely that the player can make a subtraction sentence.

Some Went Hiding *(Decomposing numbers)*

You need: Up to 10 small objects, such as marbles, pennies, beans, or rocks.

Game play: Choose a target number, such as 7. The first player shows seven objects to the next player, and then quickly hides *some* of them behind their back (without showing how many). The player then shows the rest of the items to the next player, who has to say how many "went hiding." If that player gives the right answer, it is then their turn to hide some and ask the next player to answer. If they give a wrong answer, they forfeit their turn. This game appeals best to young children.

Variation: Instead of getting a turn to hide objects, the player who answers correctly may gain points or other rewards for the right answer.

Zebra Dash Subtraction *(Subtraction within 0-5)*

You need: A six-sided die and a marker for each player. Download the game board from here:

https://www.mathmammoth.com/download/Zebra-dash-subtraction.pdf

Game play: First, each player rolls the die, and the player with the highest number starts the game. On your turn, roll the die, and move your marker that many spaces. If that square has instructions on what to do, follow those. If it has a subtraction problem, answer that. If your answer is wrong, move your marker backward one space.

Every time you come to a square with a subtraction problem, follow the same process: say the answer aloud, and if you answer wrong, move your marker back one space. And if it has other instructions, follow those.

For example, let's say you answer 5 − 2 wrong, and go backward one space. That square states you move ahead one step, so you do so, and then you answer 5 − 2 again.

The person who reaches the watering hole first is the winner.

Number Bonds in the Pond *(Number bonds)*

Note: This game is optional. It is challenging for kindergarten level, if the players don't have physical objects to help them calculate. You can make it easier by providing manipulatives/objects or instructing the child to use their fingers.

You need: A standard deck (or several) of playing cards or number cards, from 1 through one less than the target number.

Preparation: Choose a target sum for the game. If the target sum is 5, make a deck of cards consisting of numbers 1 through 4. If the target sum is 6, make a deck of numbers 1-5. And so on. Place a card with the target number face up between the players, deal two cards to each player, and spread out the rest of the cards face down, like a pond, between the players.

Game play: On your turn, take one card from the pond. Now check if any two cards in your hand add up to the target number. If so, put those cards away in your personal pile. If not, it is the next player's turn. The game ends when there are no more cards in the pond. The winner is the person with most cards in their personal pile.

Variation: Allow three cards/numbers to be added to reach the target number.

Notes: Depending on the number of players, you may need several decks of cards to make the pond. When first playing this game with the child(ren), start with 5 as the target number, and advance to target numbers of 6, 7, and onward. Playing this game several times will help the child(ren) to memorize the number bonds (the sums) associated with a particular target number.

10 Out (or *5 Out* or *6 Out, etc.*) (*Number bonds*)

Note: This game is optional as it can be challenging for kindergarten level. You can make it easier by providing manipulatives/objects or instructing the children to use their fingers.

You need: Lots of number cards from 1 to the target number.

Preparation: Choose a target number (sum) for the game, such as 10. Deal five cards to each player. Place the rest face down in a pile in the middle of the table.

Game play: On your turn, you may discard the card 10 or any two cards in your hand that add up to 10. You may also ask for one card from the player to your right (as in "Go Fish"), and if they have it, that player must give it to you. If you can discard all cards from your hand this way, you're the winner and the game ends. If not (you still have at least one card in your hand), take one card from the deck, and that ends your turn.

Variations:
* Deal more than five cards.
* Deal fewer cards if there are a lot of players or the players are very young.
* Allow players to discard *three* cards that add up to the target number.

Math Mammoth Practice Zone

Subtraction Hidden Picture Game
Solve subtraction problems within 0-5 while uncovering a hidden picture. A fun game that children really like!
https://www.mathmammoth.com/practice/mystery-picture-subtraction#min=0&max=5

"7 Up" card game
You will see seven cards dealt face up. Choose any two cards that make 10 (or your chosen sum) to discard.
https://www.mathmammoth.com/practice/seven-up

Number Bonds
This colorful online activity or game allows a child to practice the concept of number bonds. The child can either work with pictures (jewels) or with numbers.
https://www.mathmammoth.com/practice/number-bonds

Further Resources on the Internet

These resources match the topics in this chapter, and offer online practice, online games (occasionally, printable games), and interactive illustrations of math concepts. We heartily recommend you take a look. Many people love using these resources to supplement the bookwork, to illustrate a concept better, and for some fun. Enjoy!

https://links.mathmammoth.com/Kch4

Subtraction as "Taking Away"

From six cookies we take away two. There are four left.

6 − 2	=	4
Six minus two	equals	four

1. **Activity.** Show the student(s) a certain number of small items. Then take away some of them, while explaining that you are taking away or *subtracting* so many. Have the student(s) find how many are left. Then state the subtraction sentence using the words "minus" and "equals". Take turns.

2. Subtract. "Take away" the crossed-out objects by covering them with your finger. Write how many are left. (Note: Practice reading each sentence with the student using the words "minus" and "equals".)

a. $4 - 1 =$ ______

b. $5 - 3 =$ ______

c. $3 - 2 =$ ______

d. $6 - 4 =$ ______

e. $2 - 2 =$ ______

f. $7 - 5 =$ ______

3. Cross out marbles to indicate subtraction. How many are left?

a. $3 - 1 =$ 2

b. $9 - 5 =$ ______

c. $8 - 2 =$ ______

d. $4 - 2 =$ ______

e. $10 - 3 =$ ______

f. $5 - 5 =$ ______

4. Solve the problems. You can draw pictures to help you.

a. Kay had five grapes. Then a bird stole two. How many grapes does Kay have now?

b. Ian saw three bugs. Then he saw three more. How many bugs did Ian see in total?

Subtracting One

1. **Activity.** Show the the child(ren) between 6 and 10 small items. The quantity should be such that it's not easy to see at first glance how many there are. Ask the child to count the items. Then take away one of them, and ask, "How many are there now?" If the child starts counting again, ask them if they could do it without counting from 1. (Point out, if necessary, that they can simply count backward one step.) Repeat with other quantities, and take turns. Ask the child also to say the matching subtraction sentence.

2. Count the objects. Then cross out one of them. Write the *new* total on the line.

a.		________
b.		________
c.		________

3. Write the number that is **one less**.

a. 8 ______	**b.** 2 ______	**c.** 4 ______	**d.** 10 ______	**e.** 6 ______

4. Solve the problems. You can draw pictures to help you.

a. Marissa has three dolls and Chloe has five. They put their dolls together. How many dolls are there in total?

b. Greg had 10 strawberries. Then he ate one. How many strawberries does he have now?

5. Practice subtracting one.

a. 5 – 1 = ______	**b.** 3 – 1 = ______	**c.** 9 – 1 = ______
d. 7 – 1 = ______	**e.** 1 – 1 = ______	**f.** 8 – 1 = ______

How Many Are Left?

Jared had six limes: 

He gave two limes to Eva. (Cross them out.) How many does he have left?

We can write a subtraction: $6 - 2 = 4$

Six minus two equals four. Jared has four limes left.

1. Play Minus One (see the chapter introduction).

2. How many are left? Cross out objects to subtract them.

a. $7 - 6 =$ _____	**b.** $3 - 2 =$ _____	**c.** $2 - 2 =$ _____

3. Solve the problems.

a. Landon had . He gave six strawberries to Mia. How many strawberries does he have left?

b. Leila picked . She gave four flowers to her mom. How many flowers does she have now?

When we subtract zero from a number, it is the same as not subtracting anything. The number stays the same.

$7 - 0 = 7$

4. Subtract.

a. $3 - 0 =$ _____	**b.** $6 - 0 =$ _____	**c.** $10 - 1 =$ _____
d. $5 - 2 =$ _____	**e.** $6 - 1 =$ _____	**f.** $10 - 0 =$ _____

Subtracting Two

1. **Activity.** Show the the child or student(s) between 6 and 10 small items. The quantity should be such that it's not easy to see at first glance how many there are. Ask the child to count the items. Then take away *two* of them, and ask, "How many are there now?" If the child starts counting again, ask them if they could do it without counting from 1. (Point out, if necessary, that all they need to do is count backward two steps from the quantity they already counted.) Repeat with other quantities. Ask the child also to say the matching subtraction sentence, or say it yourself.

2. Count the objects. Then cross out two objects. Write the *new* total on the line.

a.		______
b.		______

3. Let's count backward! Complete the sequences.

a. 10 9 _____ _____ _____	**b.** 4 3 _____ _____ _____

4. Subtract two. You can count backward! Exercise 3 can help you, also.

a. 10 – 2 = _____	**b.** 4 – 2 = _____	**c.** 2 – 2 = _____
d. 3 – 2 = _____	**e.** 7 – 2 = _____	**f.** 8 – 2 = _____

5. Solve the problems. You can draw a picture to help you.

a. Mom needs to make six pizzas for dinner. She already made  .
How many more does she need to make?

b. Valerie had .
She gave two ducklings to Gary.
How many does she have left?

Subtraction on a Number Path 1

Let's solve 8 − 3 using the number path. We begin at 8, and we count 3 steps backward. We end up at 5.

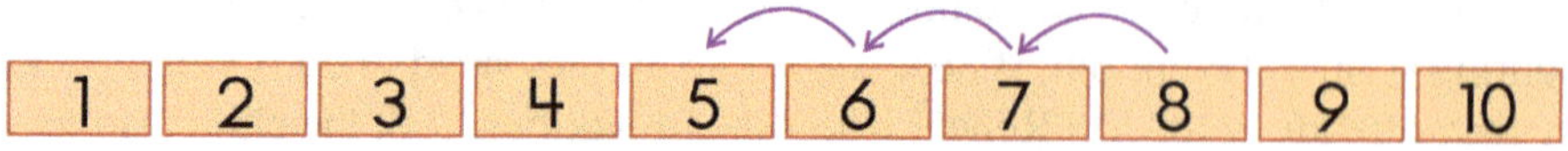

$$8 - 3 = 5$$

"Eight" "Seven, six, five."

Solve 10 – 3 using the number path. Put your finger at 10. Go three steps backward. Where do you land?

How about 8 – 4 ?

1. Use the number path above to help you solve these problems. *You can also draw a number path on a whiteboard.*

a. 7 – 4 = ______ **b.** 8 – 2 = ______ **c.** 9 – 3 = ______

d. 5 – 3 = ______ **e.** 6 – 4 = ______ **f.** 10 – 4 = ______

2. Play Minus One (or Two). See the chapter introduction. Play both versions (Minus One and Minus Two).

3. Write the subtraction sentence shown on the number path.

a. 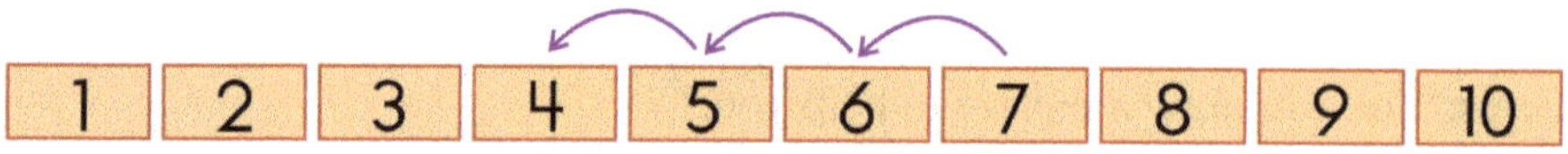

_____ – _____ = _____

b. 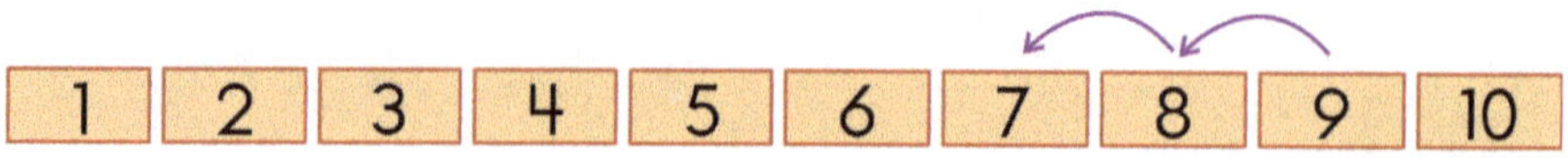

_____ – _____ = _____

4. Jeremy drew this picture to show a certain subtraction. What subtraction is he illustrating?

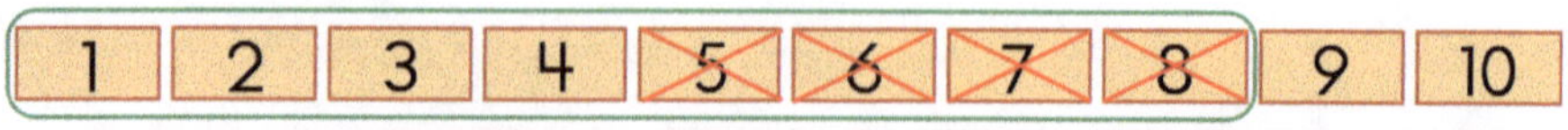

Subtraction on a Number Path 2

What subtraction is shown on the number path?
(How could *addition* be shown using arrows on a number path?)

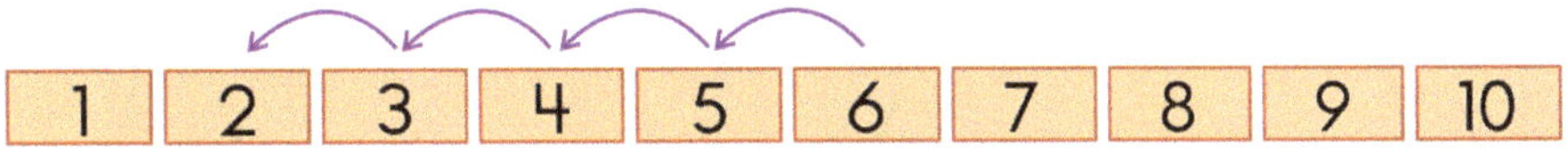

1. Add and subtract. The number path can help you. *You can also draw a number path on a white board.*

1	2	3	4	5	6	7	8	9	10

a. $6 - 3 =$ ______ **b.** $10 - 4 =$ ______ **c.** $5 + 4 =$ ______

d. $6 + 3 =$ ______ **e.** $10 - 6 =$ ______ **f.** $9 - 5 =$ ______

2. Write the subtraction sentences shown on the number paths.

a. ______ − ______ = ______

b. ______ − ______ = ______

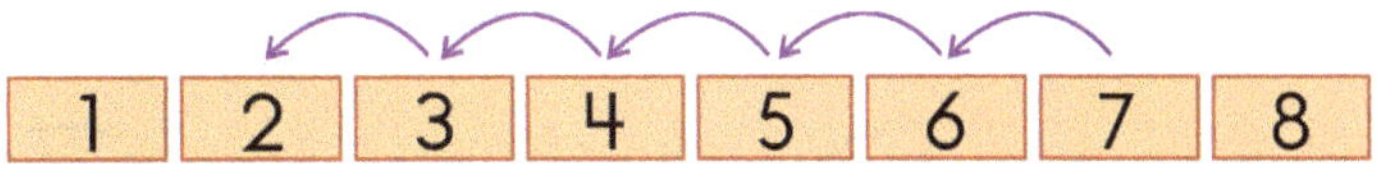

3. Solve.

a. There were six muffins in a bowl. Dad ate two of them. How many are left?

b. Natasha saw three apples in a bowl. Then Mom put  in it. How many apples are in the bowl now?

4. Tell a story about this picture using math and numbers .

Number Bonds 1

1. Count all the dots and write the total in the big circle. These are **number bonds**!

2. Draw more dots in the empty circle so that the total is the number in the big circle.

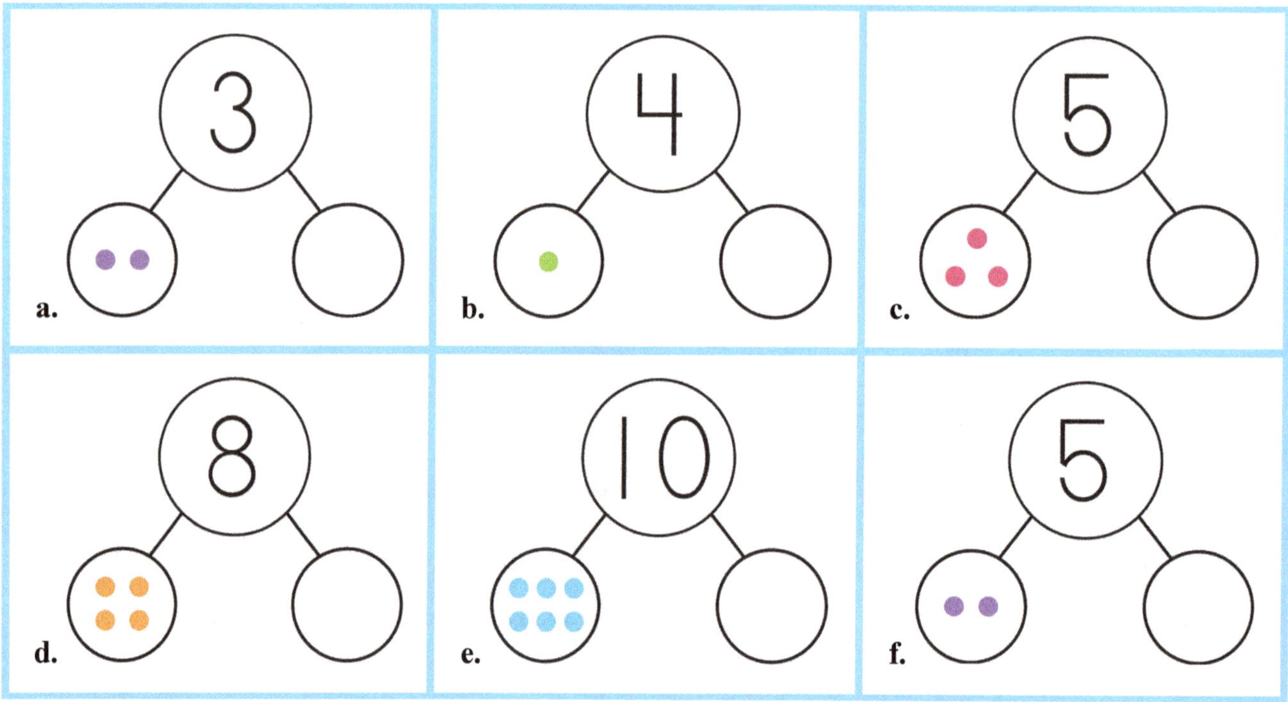

3. Match the first group of circles with the second group that makes the target number.

Number Bonds 2

1. Write the total in the big circle. (Hint: Count on from the number that is in the first circle.)

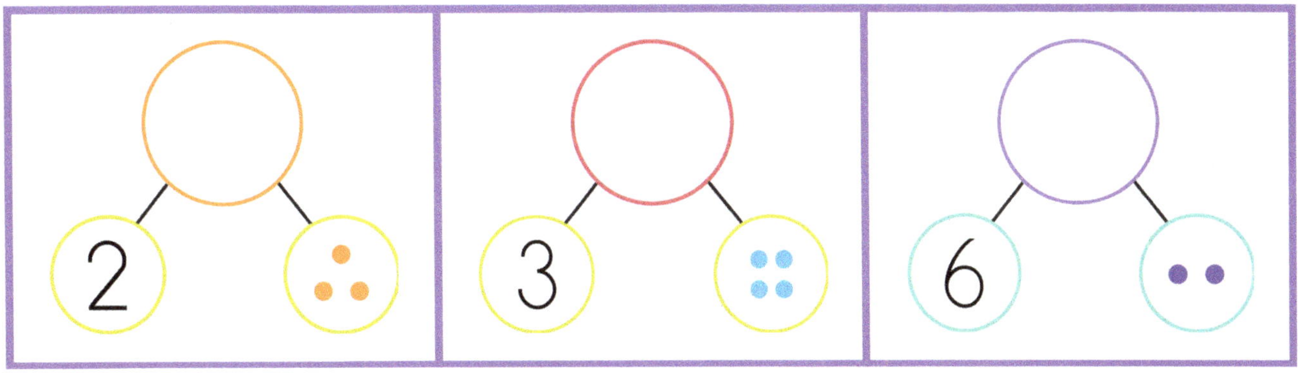

2. Draw more **dots** in the empty circle so that the total is the number in the big circle.

3. Make your own number bonds! Draw dots in the empty circles.

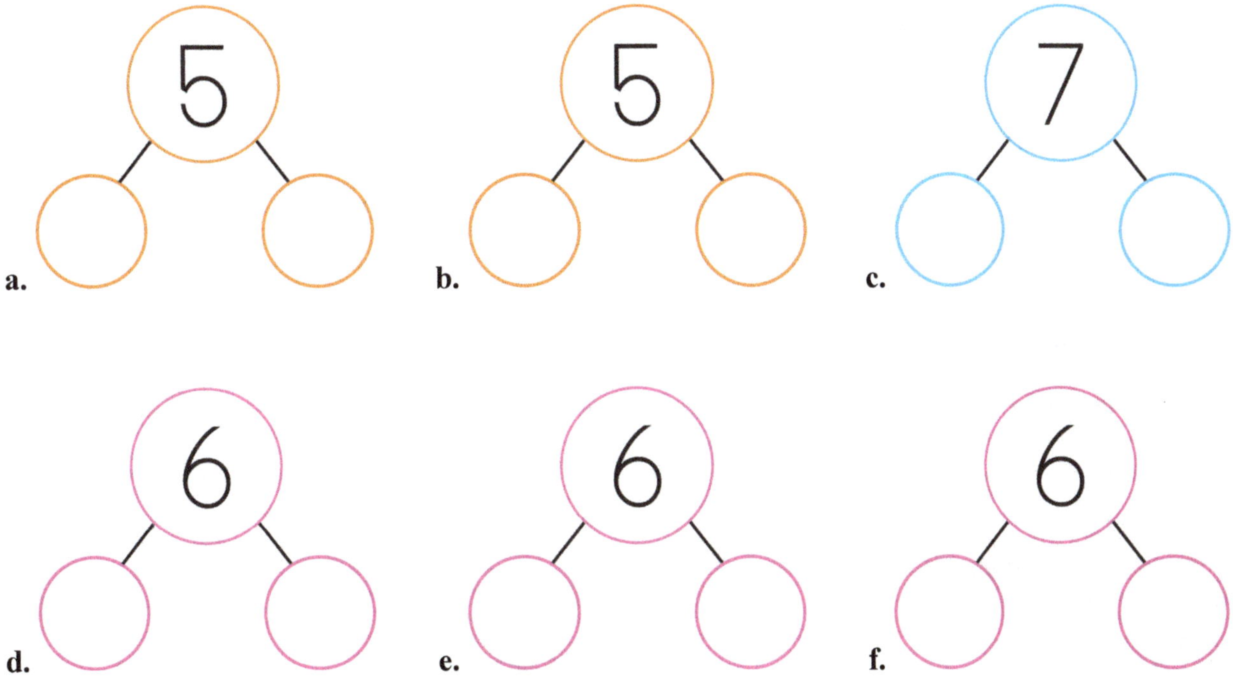

Two Parts and a Total

Here you see nine dots that are separated into two groups. We can write the addition 5 + 4 = 9.

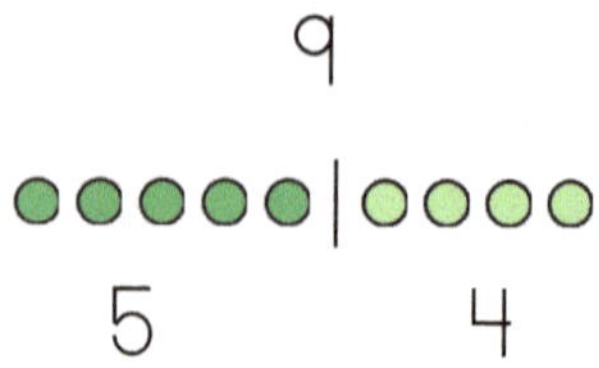

We can *also* write a subtraction to match the picture. Cover the lighter dots. Then, the darker dots are left.

What subtraction is shown by that? _____ − _____ = _____

(What if you cover the *darker* dots instead?)

1. First, write an addition. Then cover the *darker* dots and write a subtraction.

a.

_____ + _____ = _____

_____ − _____ = _____

b.

_____ + _____ = _____

_____ − _____ = _____

2. **Activity.** Make two small groups using some objects. You can choose how many are in each group. In the table below, draw your groups and write an addition and a subtraction to match your groups. Do the activity at least two times. (You can work with a partner, and also repeat it more times if desired.)

My groups:	**My groups:**
_____ + _____ = _____	_____ + _____ = _____
_____ − _____ = _____	_____ − _____ = _____

3. Draw pictures to illustrate the problem. Write an addition and subtraction sentence.
(*Note: there are two possible subtractions that can be written. The child can write either one.*)

Laura picked eight flowers. Some were purple and five were yellow. How many flowers were purple?

_____ + _____ = _____

_____ − _____ = _____

Two Parts and a Total — More Practice

1. Now we will write two subtractions. First, cover the darker flowers and write a subtraction. Then, cover the lighter flowers instead, and write another subtraction.

a. 6 − ______ = ______

6 − ______ = ______

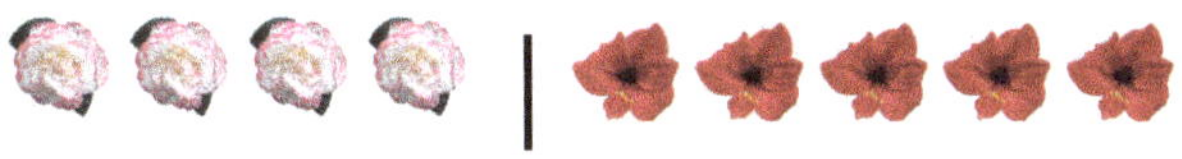

b. 9 − ______ = ______

9 − ______ = ______

2. Draw pictures to illustrate the problem. Write an addition sentence and a subtraction sentence. (*Note: there are two possible subtractions that can be written. The child can write either one.*)

Brad ate five cookies. Some were chocolate chip and two were peanut butter. How many cookies were chocolate chip?

______ + ______ = ______

______ − ______ = ______

3. Write two numbers that, when added together, make the indicated number. There are many possible ways! Then, tell your teacher two subtraction sentences that match your number bond. (You may draw dots to help you.)

a.

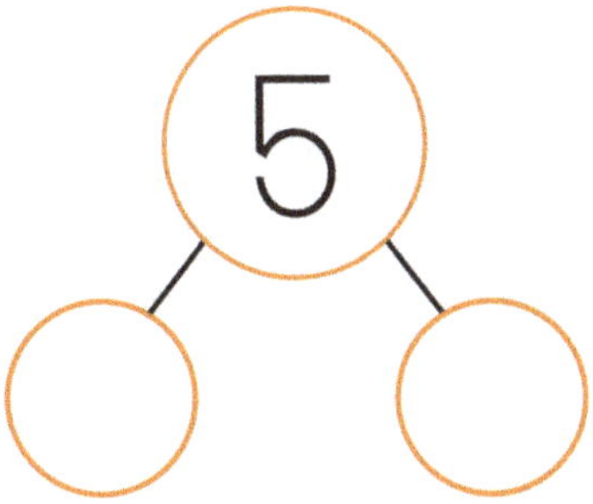

b.

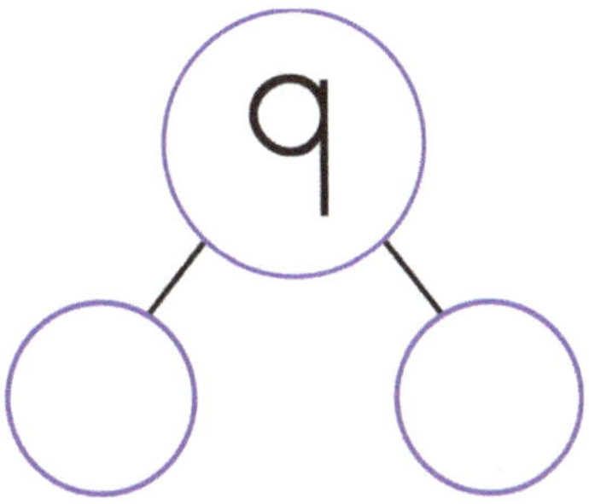

c.

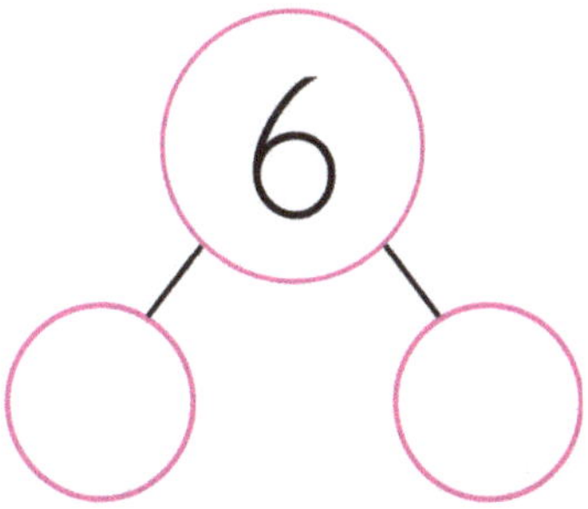

4. Continue the color pattern.

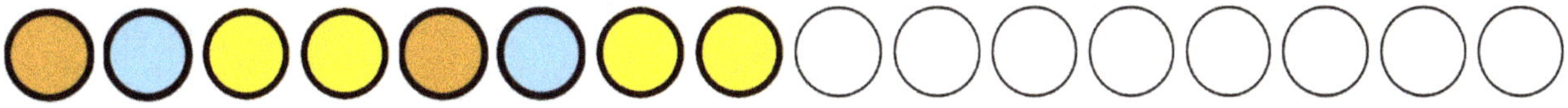

5. How many more should you add to the number in order to make 10? Use drawings or objects to find out.

a. 6 + ______ = 10	**b.** 4 + ______ = 10	**c.** 3 + ______ = 10

Subtraction within 0-5

1. "Take away" objects by crossing them out. Write how many are left.

a. $5 - 3 =$ ______

b. $3 - 2 =$ ______

c. $4 - 4 =$ ______

2. Subtract. Notice the patterns.

a. $5 - 0 =$ ______

$5 - 1 =$ ______

$5 - 2 =$ ______

b. $2 - 0 =$ ______

$2 - 1 =$ ______

$2 - 2 =$ ______

c. $4 - 0 =$ ______

$4 - 1 =$ ______

$4 - 2 =$ ______

3. Circle the number that is **less**. If the two numbers are **equal**, circle both numbers.

$4 > 1$ $7 = 7$ $0 = 0$ $5 < 9$

4. Subtract. You can use the number path to help you.

1	2	3	4	5	6	7	8	9	10

a. $4 - 3 =$ ______ **b.** $8 - 2 =$ ______ **c.** $10 - 3 =$ ______

5. Solve.

a. Kevin had four blueberries. His sister gave him some more. How many blueberries does he have now?

4

b. There were three cookies on a plate.
Jeremy ate all three cookies.
How many cookies were left?

Subtraction within 0-5, Part 2

1. Play Subtraction Challenge (see the chapter introduction). Limit the numbers to 1-5.

2. Subtract.

a.	b.	c.	d.	e.	f.
$\begin{array}{r} 5 \\ -2 \\ \hline \end{array}$	$\begin{array}{r} 3 \\ -0 \\ \hline \end{array}$	$\begin{array}{r} 2 \\ -1 \\ \hline \end{array}$	$\begin{array}{r} 4 \\ -3 \\ \hline \end{array}$	$\begin{array}{r} 3 \\ -2 \\ \hline \end{array}$	$\begin{array}{r} 1 \\ -1 \\ \hline \end{array}$

3. Solve the problem.

There were  in Sarah's garden.
Four flew away. How many are left?

4. Write a subtraction sentence to match the picture.

a. 	b. 	c. 
____ – ____ = ____	____ – ____ = ____	____ – ____ = ____

5. How many paperclips long are the objects below? Write the number on the line.

________ paperclips

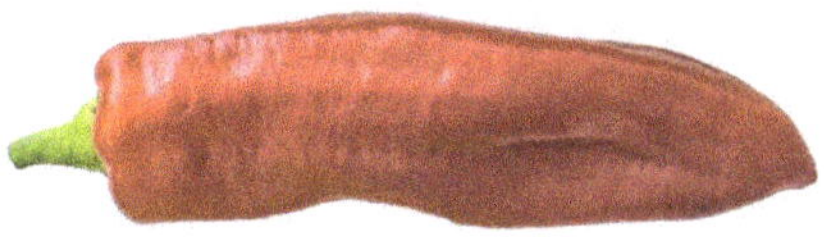

________ paperclips

6. In the space above, sketch a line that is four paperclips long.

7. Subtract.

1	2	3	4	5	6	7	8	9	10

a. $8 - 2 =$ ______ **b.** $6 - 4 =$ ______ **c.** $9 - 3 =$ ______

Subtraction within 0-5, Part 3

1. Play Subtraction Challenge (see the chapter introduction). Limit the numbers to 1-5.

2. For each subtraction problem, draw dots and cover some of them to see the answer. Then write a matching addition.

a. 5 – 3	**b.** 4 – 3	**c.** 5 – 4
____ + ____ = ____	____ + ____ = ____	____ + ____ = ____

3. For each addition problem, draw dots to illustrate it. Then write one matching subtraction.

a. 2 + 3	**b.** 4 + 2	**c.** 1 + 3
____ – ____ = ____	____ – ____ = ____	____ – ____ = ____

4. Circle the greater number.

a. 3 5 **b.** 9 6 **c.** 10 8 **d.** 4 7

5. Subtract. Notice the patterns.

a.	**b.**	**c.**
3 – 0 = ____	4 – 0 = ____	5 – 0 = ____
3 – 1 = ____	4 – 1 = ____	5 – 1 = ____
3 – 2 = ____	4 – 2 = ____	5 – 2 = ____
3 – 3 = ____	4 – 3 = ____	5 – 3 = ____
	4 – 4 = ____	5 – 4 = ____
		5 – 5 = ____

Subtract from 6

Cover two apples. What **subtraction** does that illustrate?
(Tell your teacher.)

Now imagine that the covered apples and the not-covered apples are two groups of apples. You uncover the two apples — it's like adding the two apples to the others. What **addition** can you write (or say)?

1. Subtract from six. Then write a matching addition.
   You can cover apples, and imagine adding them back in.

a. $6 - 2 =$ ____ ____ + ____ = 6	**b.** $6 - 5 =$ ____ ____ + ____ = 6
c. $6 - 6 =$ ____ ____ + ____ = 6	**d.** $6 - 3 =$ ____ ____ + ____ = 6

2. Draw more dots of a different color to make the indicated number. Complete the addition sentence.

a. ●● 2 + ____ = 4	**b.** ● 1 + ____ = 6	**c.** ●●● 3 + ____ = 5

3. Solve. You can draw something or use objects to illustrate the problem.

There are three things on the table.
Some are books and one is a cup.
How many books are on the table?

4. Subtract.

a. $5 - 5 =$ ____ **b.** $1 - 0 =$ ____ **c.** $4 - 3 =$ ____ **d.** $3 - 2 =$ ____ **e.** $6 - 0 =$ ____ **f.** $2 - 1 =$ ____

Subtract from 7

1. Play one of the subtraction games or "Some Went Hiding" with seven objects (see the chapter introduction.)

2. Count backward to zero (orally), starting from... **a.** 6 **b.** 8 **c.** 10.

3. Practice subtracting from seven. You can cover flowers to help you.

a. $7 - 7 =$ ____	**b.** $7 - 1 =$ ____	**c.** $7 - 3 =$ ____
d. $7 - 5 =$ ____	**e.** $7 - 4 =$ ____	**f.** $7 - 0 =$ ____

4. First, write an addition. Then cover the *lighter* dots and write a subtraction.

a. 	**b.** 
____ + ____ = ____	____ + ____ = ____
____ − ____ = ____	____ − ____ = ____

5. Solve. You can draw pictures to illustrate the problem.

Mom bought five pieces of fruit.
Some were pears and three were apples.
How many were pears?

6. Write two numbers that, when added, make the indicated number. You may draw dots to help you.

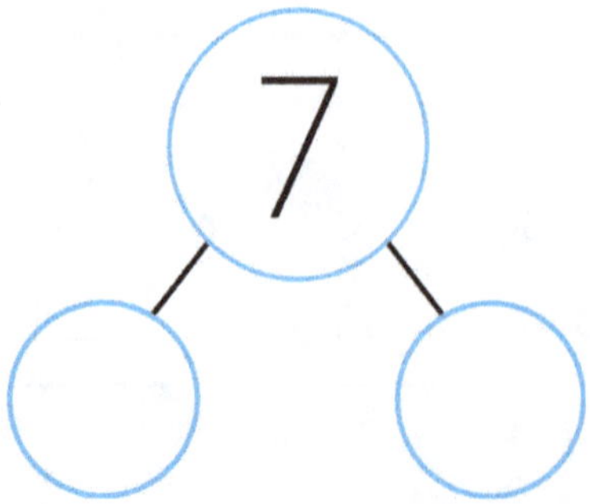

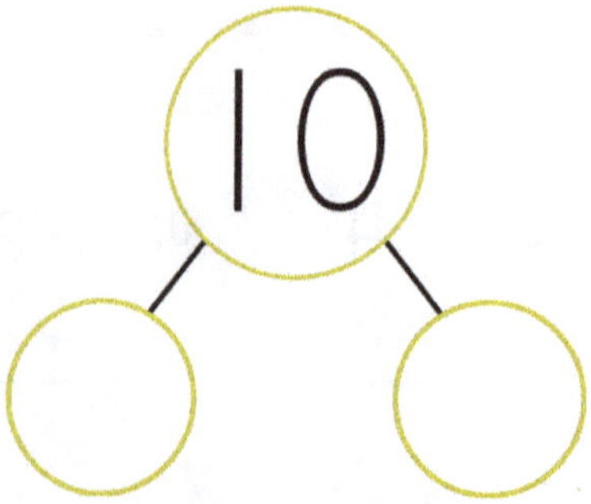

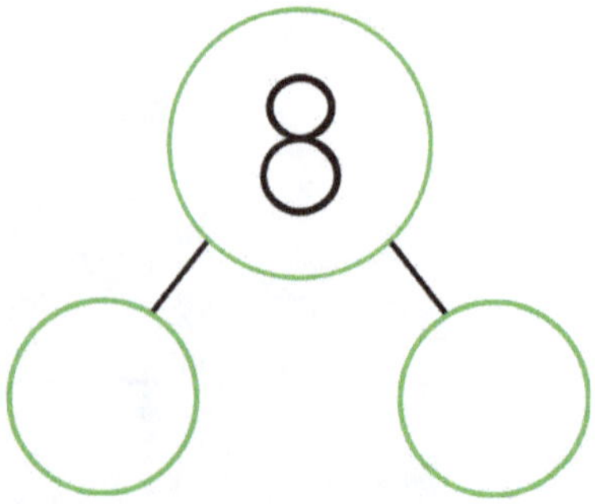

Word Problems

1. Solve. You can draw pictures or use objects to help you.

a. Seven children were jumping rope.
Three of them stopped to rest.
How many children are jumping rope now?

b. Jason had six marbles. He lost four of them.
How many marbles does he have left?

c. Jade picked one red flower, three pink flowers, and two yellow flowers. How many flowers did she pick?

2. Brandon found three seashells at the beach. Then he found some more. How many seashells did he find in total?

3. Solve. Write an addition and subtraction sentence. You can draw something to help you solve the problems.
(Note: there are two possible subtractions that can be written. The child can write either one.)

a. Connie baked seven pies. Some were lemon and five were apple. How many pies were lemon?

_____ + _____ = _____

_____ − _____ = _____

b. Luke has five pencils. Some are yellow and three are blue. How many are yellow?

_____ + _____ = _____

_____ − _____ = _____

Subtract from 8

1. Fill in. Notice the pattern in the subtractions. Then, connect each subtraction to a matching addition. (Most of the additions will get connected to two subtractions.)

8 − 0 = ____	2 + ____ = 8	8 − 4 = ____
8 − 1 = ____	4 + ____ = 8	8 − 5 = ____
8 − 2 = ____	7 + ____ = 8	8 − 6 = ____
8 − 3 = ____	5 + ____ = 8	8 − 7 = ____
8 − 4 = ____	0 + ____ = 8	8 − 8 = ____

2. Solve.

a. There are  in a box.
Two kittens jumped out.
How many kittens are in the box now?

b. There were  playing in a field.
Then three more rabbits joined them.
How many rabbits are in the field now?

3. Circle the smaller number.

a. 10 3	**b.** 9 8	**c.** 2 5	**d.** 7 4

4. Subtract.

a. $8 - 0$ **b.** $3 - 3$ **c.** $5 - 4$ **d.** $2 - 1$ **e.** $4 - 2$

Subtract from 9

1. Subtract from nine. Then write a matching addition.
   You can cover fish, and imagine adding them back in.

a. $9 - 4 =$ ____ ____ + ____ $= 9$	**b.** $9 - 7 =$ ____ ____ + ____ $= 9$
c. $9 - 6 =$ ____ ____ + ____ $= 9$	**d.** $9 - 1 =$ ____ ____ + ____ $= 9$

2. Write three different additions where the sum is 10.

a.	**b.**	**c.**
____ + ____ $= 10$	____ + ____ $= 10$	____ + ____ $= 10$

3. Draw a picture to illustrate the situation: there are three pencils, and the blue one is the longest.

4. Circle the number that is **greater**.

$6 > 5$ $\qquad$ $4 < 10$ $\qquad$ $7 > 3$ $\qquad$ $1 > 0$

5. Add and subtract. You can use an abacus, a number path, or objects to help you.

a. $9 - 4 =$ ____	**b.** $2 + 1 =$ ____	**c.** $9 - 8 =$ ____
d. $3 + 2 =$ ____	**e.** $0 + 4 =$ ____	**f.** $9 - 1 =$ ____

Subtract from 10

1. Play one of the subtraction games or "Some Went Hiding" listed in the chapter introduction.

2. Fill in. Notice the pattern in the subtractions. Then, connect each subtraction to a matching addition. (Most of the additions will get connected to two subtractions.)

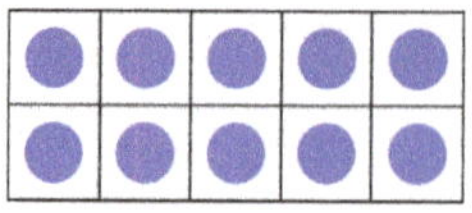

10 − 0 = _____	3 + _____ = 10	10 − 5 = _____
10 − 1 = _____	6 + _____ = 10	10 − 6 = _____
10 − 2 = _____	1 + _____ = 10	10 − 7 = _____
10 − 3 = _____	8 + _____ = 10	10 − 8 = _____
10 − 4 = _____	0 + _____ = 10	10 − 9 = _____
10 − 5 = _____	5 + _____ = 10	10 − 10 = _____

3. The total is at the top. How many are missing from the total? Draw more dots, and write the number.

a. 10	**b.** 6	**c.** 8
		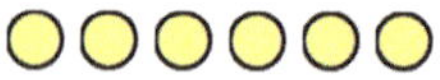
4 + _____	3 + _____	6 + _____

4. Make a word problem about these sheep for your sibling, classmate, teacher, or parent to solve.

Mixed Review Chapter 4

1. Color the picture using the color key. **Note:** The student can use objects, drawings, an abacus, *etc.* to help solve the problems.

2 = dark green	5 = red	8 = blue
3 = brown	6 = dark pink	9 = orange
4 = yellow	7 = purple	10 = green

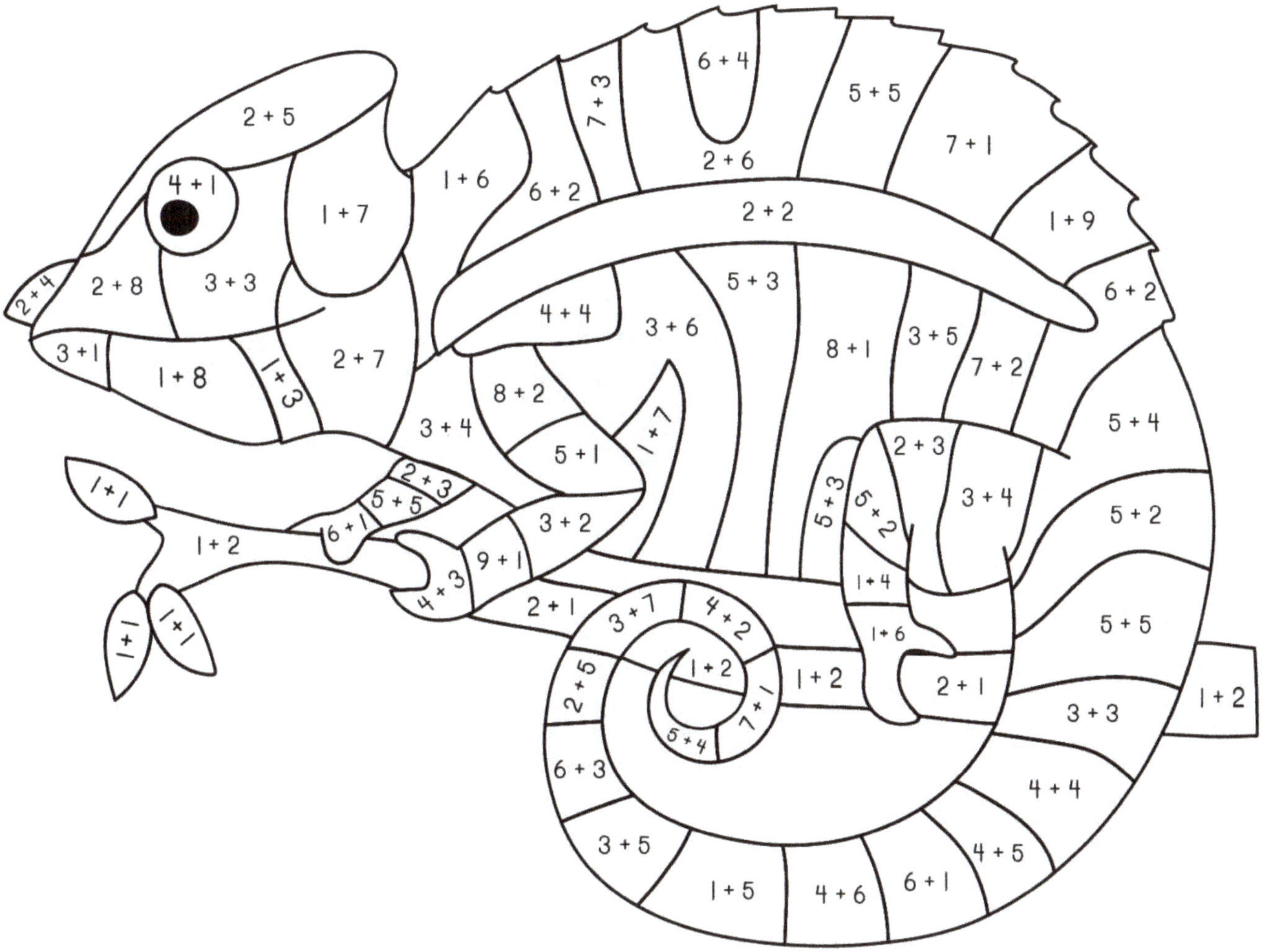

2. How many paperclips long are the objects? Write the answer on the line.

_________ paperclips

_________ paperclips

Chapter 4 Review

1. Write the subtraction sentence illustrated on the number path.

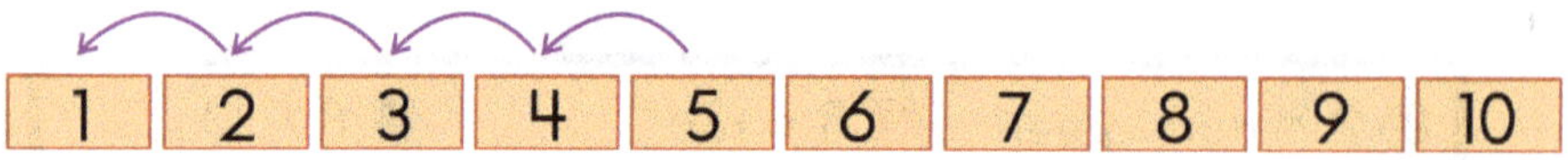

_____ − _____ = _____

2. In **a.**, write the addition that the picture shows. Then write a matching subtraction. In **b.** and **c.**, draw two groups of fruit that make the target number. Then write an addition and a subtraction.
(*Note: there are two possible subtractions that can be written. The child can write either one.*)

a.

_____ + _____ = _____

_____ − _____ = _____

b.

_____ + _____ = _____

_____ − _____ = _____

c.

_____ + _____ = _____

_____ − _____ = _____

3. Solve the problem.

Mom cut a cake into eight pieces.
The children ate 
How many pieces were left?

4. Subtract. (The student can use objects, drawings, a number path, an abacus, *etc.* to help solve the problems.)

a.	**b.**	**c.**	**d.**	**e.**
$\begin{array}{r} 9 \\ -4 \\ \hline \end{array}$	$\begin{array}{r} 5 \\ -3 \\ \hline \end{array}$	$\begin{array}{r} 3 \\ -0 \\ \hline \end{array}$	$\begin{array}{r} 10 \\ -7 \\ \hline \end{array}$	$\begin{array}{r} 7 \\ -5 \\ \hline \end{array}$

www.ingramcontent.com/pod-product-compliance
Lightning Source LLC
Chambersburg PA
CBHW061221240926
51846CB00006B/1306